BISCUIT BAKING TECHNOLOGY

BISCUIT BAKING TECHNOLOGY

Processing and Engineering Manual

SECOND EDITION

IAIN DAVIDSON
Director, Baker Pacific Ltd.

ELSEVIER

Amsterdam • Boston • Heidelberg • London
New York • Oxford • Paris • San Diego
San Francisco • Singapore • Sydney • Tokyo
Academic Press is an imprint of Elsevier

Academic Press is an imprint of Elsevier
125, London Wall, EC2Y 5AS.
525 B Street, Suite 1800, San Diego, CA 92101-4495, USA
50 Hampshire St., Cambridge, MA 02139, USA
The Boulevard, Langford Lane, Kidlington, Oxford OX5 1GB, UK

First published 2013

Second edition 2016

Notices
Knowledge and best practice in this field are constantly changing. As new research and experience broaden our understanding, changes in research methods, professional practices, or medical treatment may become necessary.

Practitioners and researchers must always rely on their own experience and knowledge in evaluating and using any information, methods, compounds, or experiments described herein. In using such information or methods they should be mindful of their own safety and the safety of others, including parties for whom they have a professional responsibility.

To the fullest extent of the law, neither the Publisher nor the authors, contributors, or editors, assume any liability for any injury and/or damage to persons or property as a matter of products liability, negligence or otherwise, or from any use or operation of any methods, products, instructions, or ideas contained in the material herein.

ISBN: 978-0-12-804211-3

British Library Cataloguing-in-Publication Data
A catalogue record for this book is available from the British Library

Library of Congress Cataloging-in-Publication Data
A catalog record for this book is available from the Library of Congress

For information on all Academic Press publications visit our website at www.elsevier.com

Typeset by MPS Limited, Chennai, India
www.adi-mps.com

Publisher: Nikki Levy
Acquisitions Editor: Nina D. Bandeira
Editorial Project Manager: Ana Claudia Garcia
Production Project Manager: Lisa Jones
Designer: Victoria Pearson

CONTENTS

BIOGRAPHY

Iain graduated from the School of Industrial Design (Engineering) at Royal College of Art in London in 1965 and joined Baker Perkins Ltd. He was Industrial Design Engineer, working in the Technical Department on the design of new biscuit, bakery and candy processing machines until 1975, gaining a thorough technical knowledge of the machines and processes.

In 1975 Iain was appointed Market Development Manager at Baker Perkins, involved in developing the Baker Perkins forward planning for new business, product development and acquisitions. In 1979 Iain became International Sales Manager with responsibility for the biscuit business in Asia and Africa.

In 1990 Iain was appointed Regional Manager Asia-Pacific for Baker Perkins and relocated to Indonesia and later in 1997 to China. His appointments included Managing Director of Baker Perkins (Hong Kong) Ltd and Director of Baker Perkins Japan KK.

In 1990 Iain negotiated an agreement with the Liaoning Foreign Trade Corporation in Dalian to establish a manufacturing facility for biscuit ovens. This was subsequently successful in manufacturing ovens under the supervision of Baker Perkins Japan KK engineers. In 2004 a manufacturing capability was established for Baker Pacific Ltd in India and later in China and Indonesia.

BAKER PACIFIC COMPANIES

Iain established PT Baker Pacific Mandiri in Indonesia in 2000. As the business outside Indonesia grew, Baker Pacific Ltd was established in Hong Kong in 2004 and is now the principal operating company, providing process technology and machinery for the biscuit, chocolate and candy industries. A team of experienced engineers and food technologists now provide Baker Pacific with a wide range of capabilities.

EXPERIENCE IN THE BISCUIT INDUSTRY

- Engineering design of biscuit process machines including a range of baking ovens
- Biscuit baking oven manufacture in China, Indonesia and India
- Sales and marketing in Europe, Asia, North America, Africa
- Project management and service.

Iain Davidson

ACKNOWLEDGEMENTS

I wish to record the contributions to our business from our team, John Lilley, Steve Eldridge, David Young, Sunny Xu, Glyn Sykes and our colleagues in the baking industry, Jean Marc Simandoux, P.K. Nandy and our colleagues over the years in the Baker Perkins Group worldwide.

Iain Davidson,
October 2015

INTRODUCTION

Our book is way of sharing over 50 years of experience in the biscuit baking industry worldwide. As engineers, we have worked with first class food technologists, bakers and confectioners in Europe, North and South America, Africa, Asia and Australasia. Our customers have included the largest multinationals and many local businesses.

You will find here a great deal of technical information and recommendations regarding biscuit baking. These are offered to share experience which has been valuable for us. It is not the only way to bake and to build biscuit baking ovens, but it has worked well for us and many of the bakeries we served.

Please note that the formulations and process information are highly dependent on the specification and properties of the local ingredients and the production equipment. All formulations and process have to be developed at the commissioning stage to suit local conditions.

My experience has taught me that it takes a small team to build the ovens. At our company a team of five people has developed the ovens described in this book: a project manager, design engineer, installation engineer, site supervisor and commissioning engineer, with a local manager to introduce and liaise with the local contractors. All manufacture has been carried out under contract by excellent engineering companies in China, India and Indonesia.

We believe that this may increasingly become a feature of future biscuit oven supply. Larger biscuit makers with substantial production facilities, experience and investment in development may lead the way to future improvements in product quality and production efficiencies. Smaller local teams may manufacture and install the ovens.

CHAPTER 1

The Biscuits

Contents

This book is primarily about the design, manufacture and operation of biscuit baking ovens. It covers in detail many aspects of the mechanical and electrical engineering of the ovens. However, the starting point of our discussion must necessarily be the biscuits and the baking process. The book is about how to bake better biscuits and to bake economically to deliver the products that consumers enjoy to eat and at an affordable cost.

Unfortunately oven builders have often given primary consideration to engineering factors such as manufacturing cost, speed of installation, ease of control and other hardware and software issues. The industry has often made cost reduction a major aspect of oven design over the last 40 years. Some of these developments have been done at the expense of the baking process and the consequent quality of the biscuits, as we shall see in the following sections of the book.

Our focus will be constantly on how to bake high-quality biscuits and the optimum oven design features to achieve this. Alongside this focus, we will also consider the design implications for efficiency of heat transfer, fuel cost, increased output and other important production considerations, but the quality of the product should remain foremost.

Let us first describe briefly the types of products our discussion will cover. Biscuits are the first and the best snack food – they need no preparation, are eaten straight from the pack, have a long shelf life, are nutritious and available in many functional forms. Biscuits are now a truly international food, consumed in large and increasing quantities in the Americas, Asia, Australasia, Africa, Middle East, as well as their original source in Europe.

Biscuits broadly fall into four categories, distinguished by their recipes and process: crackers, hard sweet and semi-sweet biscuits, short dough biscuits and cookies (including filled cookies). Each category and each product type requires a particular mixing, forming and baking process.

An example is given of the formulation and process for a product in each main biscuit category. Note that all formulations, recipes and process information given are only guides, and their use depends on the local ingredients and production equipment. In each case the recipes and process will require development during commissioning to suit the particular ingredients and equipment available.

In general the biscuit-making process follows the main steps shown below. Mixing and fermentation or dough standing time, is usually a batch process. The forming, baking, oil spraying and cooling are continuous operations with a high degree of automation. Packing is generally offline, unless the line is dedicated to a single product (Fig. 1.1).

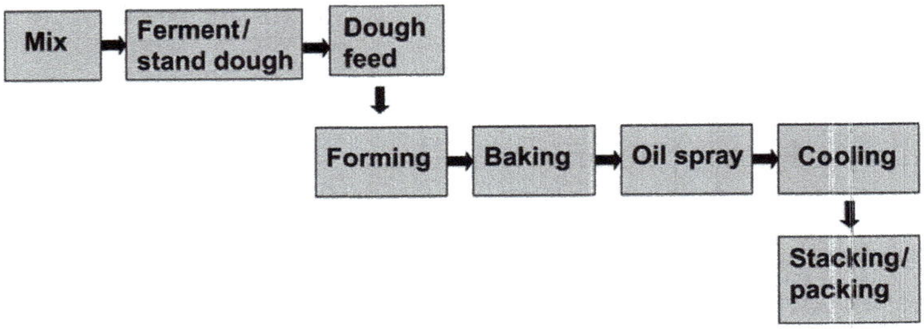

Figure 1.1 Typical process flow for crackers. Mixing and fermentation are usually batch processes, forming, baking, oil spray and cooling are continuous in-line processes.

1.1 CRACKERS

Crackers are a wide range of products characterised by crispy, open texture and savoury flavours. Crackers include soda and saltine crackers, cream crackers, snack crackers, water biscuits, puff biscuits, 'Maltkist' (sugar-topped crackers), 'TUC' type, 'Ritz' type, vegetable and calcium crackers (Fig. 1.2).

In general, crackers may have some of the following features which influence the baking process:

- Doughs which are leavened and fermented with ingredients such as yeast, ammonia and sodium bicarbonate.
- Doughs generally have a high water content (15–25%).
- Cracker doughs are laminated (the dough sheet is made up from multiple thin layers).
- Cracker doughs spring or lift in the first part of the oven to achieve an open, flaky texture; this requires humidity and high heat input.
- Some crackers are cut and baked in strips or complete sheets and broken into individual biscuits after baking.
- Some crackers require a colour contrast between dark blisters and a pale background colour.
- Traditional English crackers, such as cream crackers and water biscuits, are normally baked on light wire-mesh bands.

Figure 1.2

- Traditional American crackers, such as soda or saltine, are baked on heavy mesh oven bands which are preheated to transfer heat rapidly by conduction into the dough pieces.
- Crackers are baked to low moisture contents (1.5–2.5%), which requires a high energy input.

1.2 PROCESS FOR SNACK CRACKERS

Figure 1.3

1.2.1 Description

Snack crackers are successful in every market: light and crispy with oil spray.

1.2.2 Product Specification

Dimensions	48.0 mm diameter
Thickness	4.9 mm
Weight	3.0 g
Appearance	Evenly blistered
Colour	Golden
Texture	Light and crispy
pH	5.5
Moisture	1.3–2.5%

1.2.3 Formulation

Flour	100.00
Sugar	8.02
High-fructose corn syrup	2.85
Vegetable oil (soya bean)	11.66
Lecithin	0.20
Ammonium bicarbonate	1.84
Sodium bicarbonate	1.08
Acid calcium phosphate	1.08
Salt	0.77
Enzyme	0.01
Water	29.47

1.2.4 Critical Ingredients

1. Flour should be weak with a protein content of 8–9%
2. Proteolytic enzyme.

1.2.5 Mixing

An 'all in' mix on a horizontal mixer

Temperature of about 33°C for enzyme doughs (Fig. 1.4).

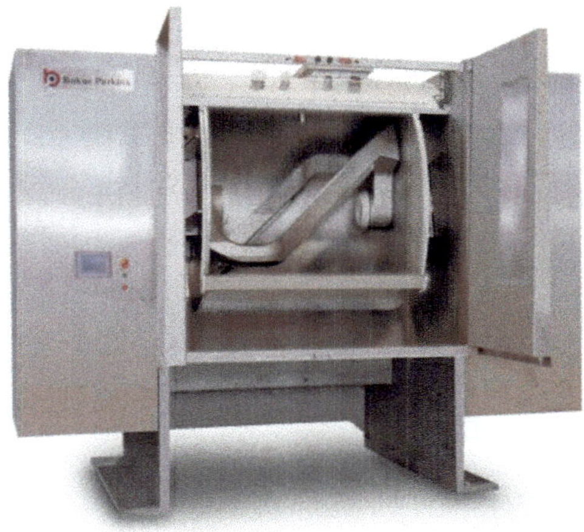

Figure 1.4 Baker Perkins horizontal high-speed mixer with shaft-less blade.

1.2.6 Standing Time

After mixing, the dough is stood to allow the enzyme to react with the gluten. The standing time is about 2.0–2.5 hours at 35°C. The time must be determined carefully depending on the amount of enzyme and the quality of the flour in order to achieve the soft, delicate eating texture of the cracker.

1.2.7 Forming

The dough is laminated with four laminations, approximately 4 mm thick

No fat/flour filling is used (Fig. 1.5).

Figure 1.5 Baker Perkins forming line: (right to left) laminator, three gauge roll units, relaxation conveyor, rotary cutter, scrap lift and return conveyors.

1.2.8 Baking

Baking time	5.0 minutes
Zone temperatures	220/220/230/230/180°C
Band preheated to	180–200°C

High bottom heat in Zones 1 and 2 of the oven is required.

1.2.9 Baking Band

Z47 type wire-mesh band

Preheat is required.

1.3 PROCESS FOR SODA CRACKERS

Figure 1.6

1.3.1 Description

Soda crackers are a traditional product in the United States, where they are made in very large volumes. Similar crackers are

'Saltine' or 'Premium' crackers. The process has several important characteristics:

- A two-stage mixing process known as 'sponge and dough'
- A long fermentation, usually 24 hours
- Fast baking time, around 2.5 minutes, on a heavy mesh pre-heated oven band
- Soda crackers are baked in strips or sheets and are broken into individual crackers (usually in pairs) after baking.

1.3.2 Product Specification

Dimensions	91 mm × 44 mm
Thickness	5.6 mm
Weight	6.25 g
Appearance	Evenly blistered
Colour	Pale creamy colour with darker blisters, evenly spaced
Texture	Open and flaky, with a crispy bite
Flavour	Mild, fermented flavour
pH	7.2–8.0
Moisture	2.5%

1.3.3 Formulation

1.3.3.1 Sponge

Flour (strong)	66.7
Fresh yeast	0.17
Dough fat	5.00
Lecithin	0.53
Malt extract 80%	0.95
Water	28.0

1.3.3.2 Dough

Sponge (follow the formulation as described in Section 1.3.3.1)

Flour (weak)	33.3
Dough fat	5.00
Soda	0.60
Salt	1.50

1.3.4 Critical Ingredients

A strong flour produced with 30% of hard wheat will give an open cracker texture. The flour used in the sponge must be 10–11% protein. Stronger flour gives a harder cracker. A weaker flour (8.0–9.0% protein) is usually used for the dough and will give a product with a softer bite.

1.3.5 Mixing and Fermentation

The sponge and dough are usually mixed on vertical spindle mixers. Two or three spindle machines are used with slow mixing speed, 25 rpm. The slow and gentle mixing action incorporates the ingredients well without undue work input at the sponge stage (Fig. 1.7).

The sponge is mixed as an 'all in' mix. The yeast should be dispersed in water before feeding to the mixing bowl. The dough is mixed to a temperature of 30–35°C, which is the optimum temperature for the action of the yeast. The sponge is fermented for 18 hours at a temperature of 30–35°C and an RH of approximately 80%. During this time the pH value will change from about 5.8 to 4.0, and the temperature of the sponge will increase.

After the fermentation of the sponge, the dough trough is taken back to the mixing room. The additional ingredients for the dough are added, including the sodium bicarbonate. Gentle, slow speed mixing is required until a homogeneous dough is made. Over-mixing will reduce the spring and give a hard, tough product. After mixing, the dough is returned to the fermentation room for up to 6 hours. With the addition of soda, a large change in the pH occurs and the dough will reach a pH of over 7.0.

1.3.6 Dough Forming

The dough is laminated, usually with four to six layers at around 4.0 mm thickness. The dough is then gauged with a maximum reduction at each gauge roll unit of 2:1. Excessive reductions

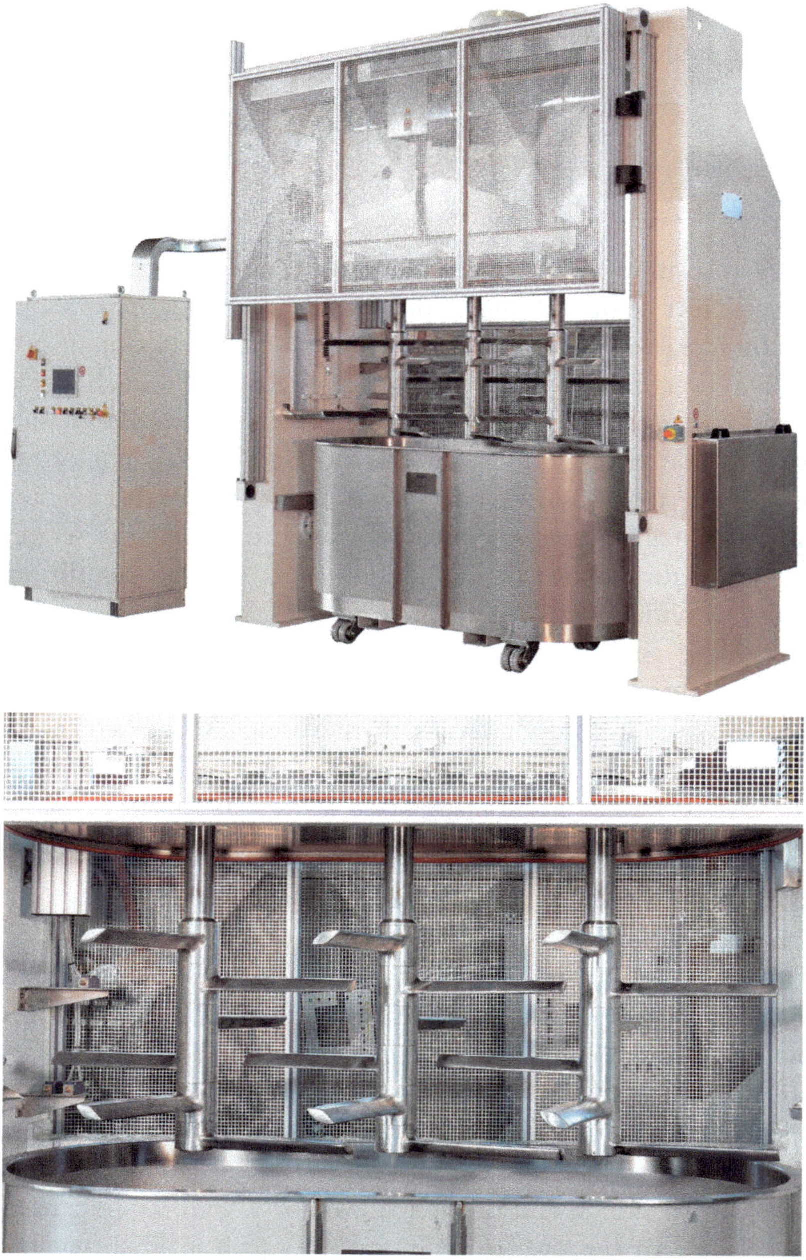

Figure 1.7 Vertical spindle mixer from Apinox Srl. These pictures belong to Apinox Srl, Italy. This mixer has been specifically designed for the production of soda cracker, cream cracker with two stages, fermented dough and delicate dough.

of the dough thickness will prevent good lift or spring of the cracker. Typical settings for the gauge roll gaps are:

Gauge roll 1	12 mm
Gauge roll 2	6 mm
Gauge roll 3	3 mm

A relaxation conveyor is used to relax the tension in the dough sheet before cutting, as the soda cracker doughs are subject to considerable shrinkage after cutting and during baking.

The dough sheet is cut with a 'scrap-less' cutter. Each cracker shape is perforated (not cut through), so that the dough sheet remains complete. A small amount of edge scrap is cut off and this is diverted by side scrap wheels to the scrap return system. An additional cross scrap conveyor is used to convey the scrap to the side return conveyor. The dough sheet may be cut through across its width with one revolution of a large diameter cutting roll so that the dough sheet is divided into lengths of approximately 1.0 m in the oven. This allows shrinkage to occur during baking without random breaks in the dough sheet, which may cause problems at the cracker breaker.

1.3.7 Baking

The baking of soda crackers normally follows the US practice with a direct gas-fired oven and heavy mesh oven band (20.5 kg/m^2). Typically, a Compound Balanced CB5 band is used with preheat to give a high band temperature, over 150°C at the feed end of the oven. Heat is immediately conducted into the bottom of the dough sheet, initiating a fast and strong lift or spring to give the open, flaky texture of the cracker.

A considerable amount of water must be evaporated from the soda cracker dough and this is achieved by a high temperature in the middle of the oven (minimum 300°C on a direct gas-fired oven). The fast baking time and high water evaporation require a powerful oven. Soda cracker ovens have a burner capacity of over

$30\,kW/m^2$ of oven band area. The first zone will have a burner capacity of $45–50\,kW/m^2$ of oven band area.

Temperature profile: 300/300/280/250°C

Baking time: 2.5–3.0 minutes.

1.3.8 Alternative Oven Band

Good-quality soda crackers can also be baked on a Z47 wire-mesh band with preheat. The process details are similar, but the baking time is longer, usually 4.0 minutes.

1.3.9 Oil Spray

Soda crackers are oil sprayed immediately after baking. The cracker strips are transferred from the oven stripping conveyor to the oil spray machine, where a mist of fine oil is sprayed on the top and bottom of the crackers. Coconut oil or palm kernel oil is used.

1.3.10 Cracker Breaking

The baked sheet is broken into lateral strips by a weighted roller positioned over the oven stripping conveyor. The wire-mesh conveyor is supported by rollers before and after the cracker breaker roll. The sheet of crackers is depressed by the breaker roll and breaks into separate strips (Fig. 1.8).

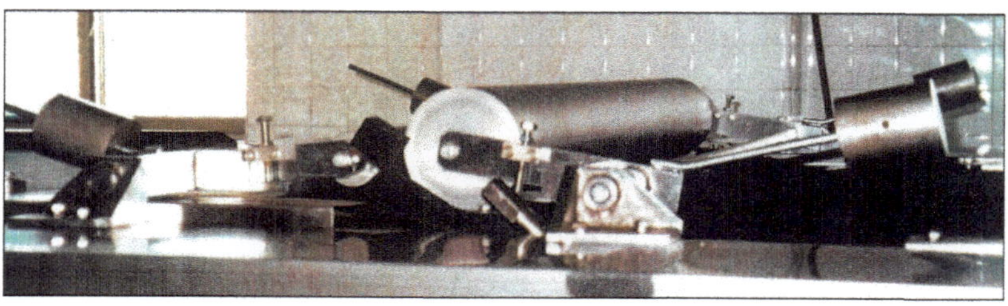

Figure 1.8 Cracker-breaking roll at the oven end.

After cooling, the cracker strips are 'penny stacked' by differential belt speeds between the final cooling conveyor and the packing table. After stacking, the strips must be correctly aligned. Usually, side guide rolls or belts are used to nudge the strips into a central position for breaking. The strips are depressed by breaker wheels aligned with the perforations in the cracker sheet and break at the perforations. The products are normally broken into pairs.

1.4 HARD SWEET BISCUITS

Figure 1.9

Examples of hard sweet biscuits are Marie, Petit Buerre, Rich Tea, Arrowroot and Morning Coffee. They are characterised by an even, attractive colour and texture and good volume.

Doughs for hard sweet biscuits have the following features:

- Doughs have strong, developed gluten which gives an elastic dough, which is sheeted and cut. It often shrinks in the first stage of baking
- Doughs have low sugar and fat
- Doughs have water contents typically of around 12%
- Biscuits are normally baked on a wire-mesh band (except for Marie which is traditionally baked on a steel band)
- Humidity in the first part of the baking is important to achieve good volume and a smooth surface sheen
- Biscuits are baked to low moisture contents, around 1.5–3.0%.

1.5 PROCESS FOR HARD SWEET BISCUITS

Figure 1.10

1.5.1 Description

Marie is a classic biscuit made throughout Europe and Asia. It has a light, crisp, delicate texture, with pale colour and clear smooth surface.

1.5.2 Product Specification

Dimensions	66.0 mm diameter
Thickness	6.0 mm
Weight	8.3 g
Appearance	Smooth surface, clear printing
Colour	Pale golden
Texture	Crisp and light
Moisture	1.5%

Marie biscuits are made with medium protein flour and contain sodium metabisulphate (SMS) to develop a soft extensible dough. The doughs are mixed on horizontal mixers to a temperature of 40–42°C.

The dough is sheeted and cut and is traditionally baked on a steel band.

1.5.3 Formulation

Flour	100.00
Cornflour	4.41
Maize flour	14.70
Granulated sugar	25.59
Invert syrup 80%	7.94
Shortening	11.03
Lecithin	0.57
Salt	0.88
Soda	0.67
Acid calcium phosphate	0.08
Protease	0.02
SMS 10% solution	0.02
Ammonium bicarbonate	0.73
Water	26.47

1.5.4 Critical Ingredients

1. Flour should not exceed 9.0% protein. Higher protein will result in a hard biscuit

2. Cornflour and maize flour are used to reduce the total gluten content and make a more tender eating biscuit
3. SMS will modify the protein to make a soft extensible dough.

1.5.5 Mixing

An 'all in mix' on a horizontal mixer.

Mixing is critical to developing the soft extensible dough. A mixing action which kneads the dough without too much tearing and extruding is ideal. Mixing time on a typical high-speed mixer will be 20–25 minutes. Marie doughs are mixed until the required temperature is achieved. The dough should reach 40–42°C. At this temperature, it should be well kneaded and of correct consistency for machining. Higher dough temperatures result in unstable doughs. The dough is used straight away without standing, and it is important to maintain the temperature.

1.5.6 Forming

The dough may be laminated, but doughs made with SMS are usually sheeted without lamination. Dough scrap incorporation is very important and should be very even and consistent. The temperature of the scrap dough should be as close as possible to the temperature of the new dough. Dough sheet reduction should be gentle and should not exceed the ratio of 2.5:1.

Typical roll gaps are:

Forcing roll gap on sheeter	18.0 mm
Gauging gap on sheeter	9.0 mm
Gauge roll 1	5.7 mm
Gauge roll 2	2.5 mm
Gauge roll 3	1.1 mm (cutting thickness: 1.3 mm)

The doughs shrink and require good relaxation before cutting

Separate cutting and printing rolls on the rotary cutter are recommended to achieve good, clear printing and docker holes (piercing of holes in the dough pieces) (Fig. 1.11).

Figure 1.11 Baker Perkins rotary cutting machine with scrap dough lifting and return conveyor.

1.5.7 Baking

Steam may be used at the oven entry to achieve a high humidity. This will improve the surface finish of the biscuit.

Baking time	5.0–6.5 minutes
Temperatures	200/220/180°C
Moisture	1.5%

1.5.8 Cooling

A ratio of cooling to baking time should be at least 1.5:1. This will help to avoid checking (cracking of the biscuits after packaging due to an internal moisture gradient).

1.6 SHORT DOUGHS: ROTARY MOULDED BISCUITS

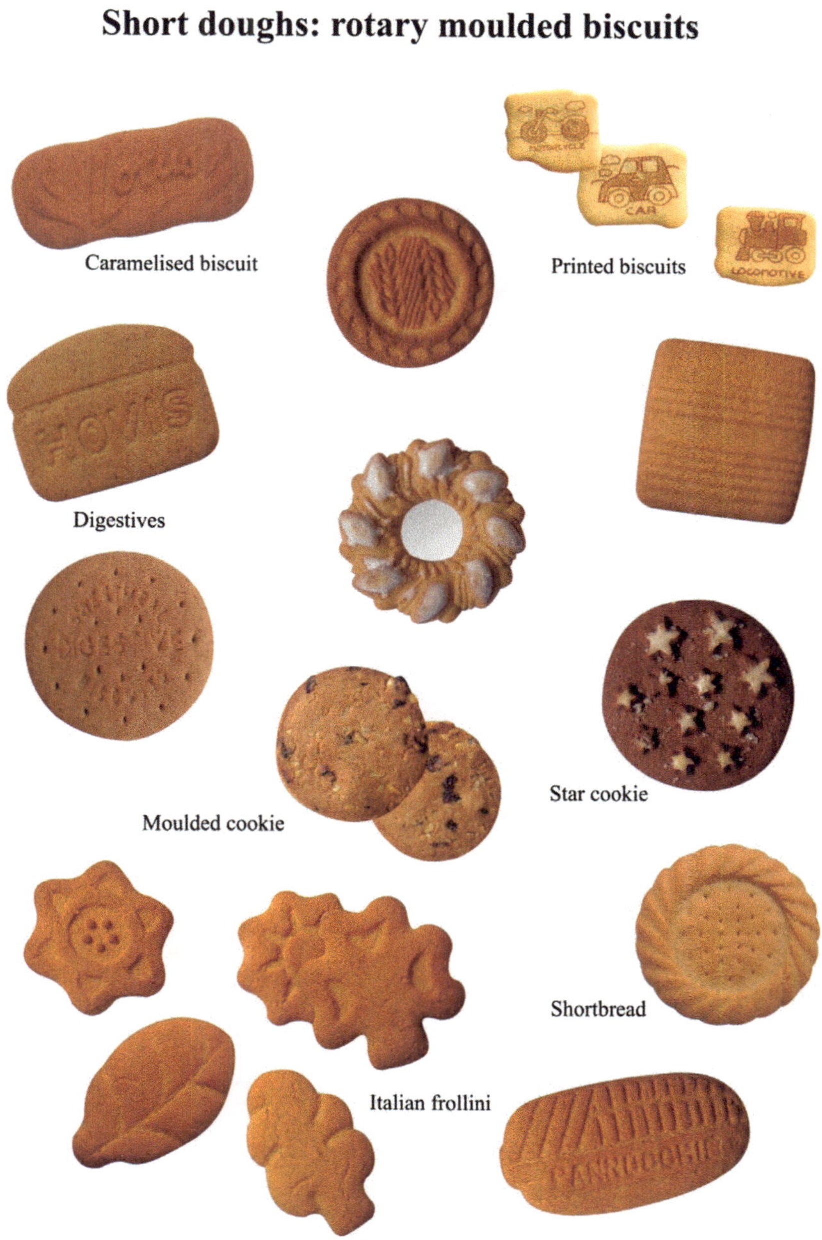

Figure 1.12

Figure 1.13

1.7 SHORT DOUGHS (MOULDED BISCUITS)

This is another wide biscuit category with many designs. The doughs are short with higher fat and sugar contents than the crackers and hard sweet biscuits. This is the simplest category for the dough piece forming process (rotary moulding) and so these products are very widely produced, often in very high volumes. The list of typical products would be very extensive and contains many local or regional specialities:

Malted milk, Glucose, Lincoln, Digestive, Nice, Shortbread, custard cream from Britain, Italian frollini, Dutch speculaas, caramelised biscuits, and Glucose and Tiger from India.

- Doughs have a low water content, but more fat and sugar than the hard sweet biscuits
- High humidity in the first part of the baking process allows the biscuit structure to form
- Relatively slow baking at comparatively low temperatures.

1.8 PROCESS FOR A MOULDED SHORT DOUGH BISCUIT

Figure 1.14

1.8.1 Description

The moulded short dough biscuit is a rotary moulded biscuit derived from 'Glucose', a highly popular biscuit in India. One maker in India, Parle Products, sells approximately 13 billion of their Parle G Glucose biscuit every month. 'Glucose' type biscuits are made in very large volumes in India and it was successfully introduced into Malaysia and Indonesia. In Indonesia, it quickly became one of the biggest selling biscuits.

The biscuit is a short, sweet rotary moulded product, developed from the English 'Malted Milk' biscuit. It is cheap, nutritious and satisfying. It is often fortified with vitamins and minerals, particularly calcium and iron.

1.8.2 Product Specification

Dimensions	58 mm × 37 mm
Thickness	6.7 mm
Weight	5.2 g
Appearance	Bold design
Colour	Golden
Texture	Short and tender
Flavour	Sweet
Moisture	1.2–1.4%

1.8.3 Formulation

Flour (weak)	100.00
Powdered sugar	29.50
Palm oil	19.50
Lecithin	0.62
Glucose 42DE	2.50
Fructose	1.25
Skimmed milk powder	2.35
Calcium carbonate	0.55
Salt	1.10
Sodium aluminium phosphate	0.30
Ammonium bicarbonate	0.58
Soda	0.49
Vitamin mix	0.11
Flavours	0.21
Water	13.33

1.8.4 Critical Ingredients

1. Flour should not exceed 8.0–9.0% protein. Higher protein will result in a tough biscuit, particularly if it is overmixed
2. The vitamin and mineral mix should be developed to suit the local requirement.

1.8.5 Mixing

The mixing process must not develop the gluten in the flour, as this will result in a tough biscuit. Mixing is therefore done in two stages.

In the first stage, all the ingredients are mixed except the flour and soda. The mixing continues until a consistent, homogenous cream is produced. It is important to keep the temperature of the dough low and chilled water is required for the mixer jacket.

For the second stage, the flour and soda are added and mixed for 1 minute at slow speed and 1–2 minutes at high speed. The final dough temperature should be 18–22°C.

1.8.6 Standing Time

The dough will be sticky when discharged from the mixer as the flour has not fully absorbed the water. The dough should be stood in a cool area for 30 minutes before forming. It will then be less sticky and will release from the rotary moulding roll more easily.

1.8.7 Rotary Moulding

The dough should be fed very evenly and consistently to the hopper of the rotary moulder, maintaining an even level across the width of the machine during the production.

Adjustment will be made to the forcing roll gap, knife position and pressure roll to achieve a good release and good product shape (Fig. 1.15).

1.8.8 Baking

Steam may be used at the oven entry to achieve a high humidity. This will allow the biscuit to expand in the first zone and achieve good volume.

Baking time	5.0–5.5 minutes
Temperatures	180/200/220/200/180°C
Moisture	<3.0%

1.8.9 Cooling

Sufficient cooling is required to set the biscuit, which will be soft as it leaves the oven, usually 1:1.5 baking to cooling time.

Figure 1.15 Rotary moulder from Dingson Food Machinery.

1.9 COOKIES

Chocolate chip cookies, butter cookies, two-dough cookies, centre-filled cookies, fig bars, fruit bars, extruded cookies, cookies with many types of inclusion such as nuts, raisins, coconut and chocolate chips.

- Very soft doughs which are deposited directly onto the oven band
- High fat and sugar recipes
- Long baking times with relatively low baking temperatures
- All products are baked on steel bands
- High humidity is required in the first oven zones to allow the dough to spread on the oven band (Figs 1.16–1.18).

Cookies

Figure 1.16

Filled cookies
Apple-filled cookie
Jam filled cookies
Chocolate-and cream-
filled biscuits
Jam tarts
Cookies with
strawberry filling
Soft centre-filled cookies
Filled bars
Soft centre-filled cookies
Crostatini
Raisin biscuit
Fig bars

Figure 1.17

Danish butter cookies

Figure 1.18

1.10 PROCESS FOR A CHOCOLATE CHIP COOKIE

Figure 1.19

1.10.1 Description

Short cookies with inclusions of chocolate chips or nuts.

1.10.2 Product Specification

Dimensions	55 mm diameter
Thickness	12.0 mm
Weight	15.0 g
Appearance	Round, irregular shape with chips visible

Colour	Golden brown
Texture	Short
Flavour	Rich with chocolate or nut flavour
Moisture	2.5–3.0%

1.10.3 Formulation

Flour	100.00
Shortening	55.98
Granulated sugar	50.05
Brown sugar	0.76
Whole egg powder	1.24
Vanillin	0.10
Invert syrup	1.24
Salt	0.96
Ammonium bicarbonate	0.29
Sodium bicarbonate	0.67
Chocolate chips	30.00
Water	19.14

1.10.4 Mixing

The mixing is in two stages on a horizontal or a vertical mixer. Good dispersion of the fat over the flour particles is important, and there must be a good proportion of solid fat.

The following ingredients are mixed gently at the first stage: shortening, sugars, water, salt, egg powder, vanilla, invert syrup and ammonium bicarbonate. These ingredients are mixed to dissolve the sugar and achieve a creamy emulsion. The water should be cold and the mix kept as cool as possible.

The flour and sodium bicarbonate are added for the second stage. The mixing is continued at low speed for no more than 1 minute to obtain a homogenous mixture without hydration of the flour and formation of the gluten. The chocolate chips or nuts are added close to the end of the mix and given enough time to disperse evenly through the dough (Fig. 1.20).

Figure 1.20 Single Sigma Arm Mixer from Peerless Food Equipment for cookie and biscuit doughs.

1.10.5 Forming

The dough is fed to the hopper of a wire-cut machine. The dough may be fed from a bowl by gravity. The feed rolls of the depositor operate continuously and will extrude the dough through the dies. As the dough is extruded, it is cut by a horizontally reciprocating wire. The cut dough pieces drop directly on to the baking tray or oven baking band (Figs 1.21 and 1.22).

1.10.6 Baking

Baking time	7.0 minutes
Temperatures	180–220°C
Moisture	2.5–3.0%

1.10.7 Cooling

A ratio of cooling to baking time should be 1:1.5.

Figure 1.21 Baker Perkins wire-cut machine.

Figure 1.22 Dough pieces being cut and deposited on to the oven band.

1.11 LONG SHELF LIFE CAKES, SNACK CAKES

A variety of snack cakes are produced on continuous tunnel ovens. These include layer cakes, baked in a complete continuous sheet, which is subsequently cooled, slit and cut into individual products; snack cakes baked individually in tins or pans; and Korean 'pies' which are soft deposited doughs, baked on a steel band.

- Cakes are produced from soft batters with relatively low viscosity.
- Some cakes (layer cakes) are baked in complete sheets of batter deposited on steel bands.
- Some snack cakes are baked in pans which are carried through the oven on chain tracks.
- Korean 'pies' are baked on steel bands as deposited cookies (Figs 1.23 and 1.24).

Note: For additional information on ingredients for biscuits, please refer to Appendix 1.

Figure 1.23

Snack cakes
Jaffa cakes
Muffin
Sandwich cake
Snack cake with injected filling
Jaffa cakes
Cherry cake
Custard cake
Saccotino
Snack cake with injected filling

Figure 1.24

1.12 SUMMARY

We have listed the main product types which can be baked on a continuous 'tunnel' oven:

- Crackers
- Hard sweet biscuits
- Soft dough biscuits
- Cookies
- Snack cakes

It will be seen that biscuit baking ovens must meet a wide range of process requirements to produce the structure, texture, volume, appearance, colour and moisture content required by each product. The ability to control heat transfer, baking temperature profile, time and humidity are critical factors in producing good-quality products. A fundamental understanding of these factors is the foundation for good biscuit oven design.

BIBLIOGRAPHY

Almond, N., 1989. Biscuits, Cookies and Crackers. Volume 2, The Biscuit Making Process. Elsevier Science Publishers Ltd, London.

Almond, N., Gordon, M., Reardon, P., Wade, P., 1991. Biscuits, Cookies and Crackers. Volume 3, Composite Products. Elsevier Science Publishers Ltd, London.

Benedict, M., 2001. How does temperature affect yeast activity? MadSci Network. <http://www.madsci.org/posts/archives/2001-01/980908832.Gb.r.html>.

Buck, J.S., Walker, C.E., 1988. Sugar and sucrose ester effects on maize and wheat starch gelatinization patterns by differential scanning calorimeter. Starch 40 (9), 353–356. Wiley VCH Verlag GmbH.

Dakota Baker's Yeast, 2010. Yeast fermentation in baked goods. <http://www.dakotayeast.com/help-fermentation.html>.

Eyre, C., 2008. AB Enzymes launch targets improved biscuit baking. Bakeryandsnacks.com. Decision News Media. <http://www.bakeryandsnacks.com/Ingredients/AB-Enzymes-launch-targets-improved-biscuit-baking>.

Fineli®, 2003–2010. National Institute for Health and Welfare. Wheat flour whole grain.

Flour specifications. <www.flourmilling.co.uk>.

Ghiasi, K., Hoseney, R.C., Varriano-Marston, E., 1981. Effects of flour components and dough ingredients on starch gelatinization. Cereal Chem. 60 (1), 58–61.

Gurney, A., 2008. Bakery Fats and Oils. Leading Edge. Leading Edge Bakery & Food Service Journal Pty Ltd., 3197 Patterson Lakes, Victoria, Australia.

Lansbergen, G., 2002. Fats for Food Consultants. DTKFCPL. Examining the Role of Fats in Bakery. APR03. <http://www.fatsforfoods.com/specifications.htm>.

Learning, Food Resource, 1990. Oregon State University. Gel. Excerpts from Bender Arnold E. Dictionary of Nutrition and Food Technology, Butterworths, Boston.

Learning, Food Resource, 2010a. Starch. Oregon State University. <http://people.oregonstate.edu/~calverta/learn/starch.html>.

Learning, Food Resource, 2010b. Bread. Oregon State University. <http://people.oregonstate.edu/~calverta/learn/bread.html>.

Lowe, B., 2010. Experimental Cookery from the Chemical and Physical Standpoint, 2nd Edition. John Wiley & Sons, New York, Chapman & Hall, London.

Manley, D., 1996. Manley's Technology of Biscuits, Crackers and Cookies, fourth ed. Woodhead Publishing Ltd, ISBN 1 85573 280 7.

Manley, D., 2001. Biscuit, Cracker and Cookie Recipes for the Food Industry. Woodhead Publishing Ltd., ISBN 1 85573 543 1.

Moodie, P., 2001. Traditional baking enzymes – proteases. In: Enzyme Development Corporation, pp. 1–10. [online]. Presented at American Institute of Baking, Manhattan, Kansas.

Peerless Food Equipment. <www.peerlessfood.com>.

Penton Media Inc., 2008. A tropical touch, soya-based solutions. <www.baking-management.com>.

Sykes, G.B., 2005–2014. Baker Pacific Ltd.

Toufeili, I., Ismail, B., Shadarevian, S., Baalbaki, R., Khatar, B.S., Bell, A.E., et al., 1999. The role of gluten proteins in the baking of arabic bread. J. Cereal Sci. 30, 255–265.

Wade, P., 1988. Biscuits, Cookies and Crackers, Volume 1 The Principles of the Craft. Elsevier Science Publishers Ltd, London.

CHAPTER 2

Baking Process

Contents

Our focus is on the baking of good-quality biscuits. The starting point must be an understanding of the ingredients and the baking process. Section 2.1 outlines the process, the changes that take place from the dough piece to the biscuit during baking and the factors that influence the baking process and the quality of the end product.

2.1 FROM THE DOUGH PIECE TO THE BISCUIT

There are three main changes which we will see as all biscuits are baked. They are the development of the biscuit structure and texture, the reduction in the moisture content and the development of the colour. These changes overlap during the baking process, but it is useful to note that the formation of the structure and texture of the biscuit will take place in the first half of the oven, the reduction in the moisture mainly in the middle of the oven and the development of the colour in the final third of the oven (Figure 2.1).

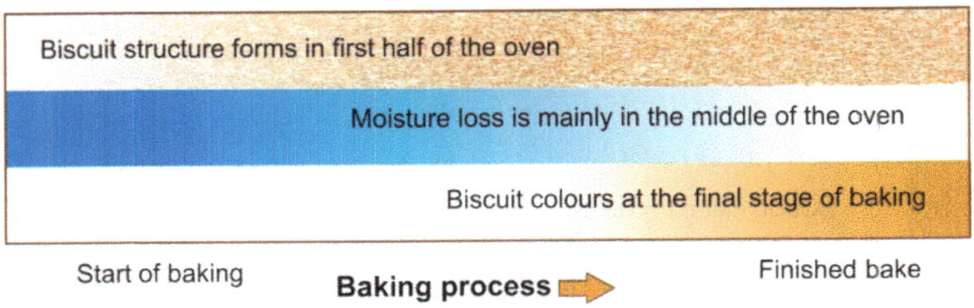

Figure 2.1

2.1.1 Biscuit Structure

Our aim is to bake a high-quality biscuit. The following characteristics are important:

- Texture – open, flaky, short, depending on the product, etc.
- Density/volume – low density gives more volume and a lighter bite
- Bite/mouth feel – crispiness, softness, smoothness, crunchiness, etc.
- Flavour – many flavours and fillings are heat susceptible and the protection of the flavours and texture of the fillings needs consideration for the baking process. For example, for a variety of soft doughs and cookies, a preference will be given for radiant heat, a longer baking time at a lower temperature (Figure 2.2).

Figure 2.2 Wide variety of biscuit textures, densities, bites and flavours.

2.1.2 Moisture Content

- An important factor in baking doughs with high water content, such as crackers and water biscuits

- Low moisture content enhances the keeping qualities of the biscuit
- Evenness of the moisture content from the centre to the outside of the biscuit requires penetrative heat and adequate time for baking and cooling to avoid 'checking' (cracks in the biscuits after packing).

2.1.3 Colour

- Consistency of colour with time and across the width of the oven band
- Some products such as Marie require a very even bland colour; others such as cream crackers and some rotary moulded designs require colour contrasts and highlights. These features require different baking systems to enhance the appearance of the product (Figure 2.3).

Figure 2.3 Variety of biscuit colours and highlighting.

2.2 INGREDIENTS

The structure and texture of the biscuit is determined by the ingredients, mixing and forming and the baking process. Here we will introduce briefly the main ingredients and process requirements for the baking of biscuits. Our aim is to indicate the complex chemical and physical changes which take place with temperature and particularly during baking. This will inform how we design and operate the baking oven.

2.2.1 Wheat Flour

The principle ingredient of biscuits is wheat flour. The grain consists of bran (12%), which is the outer husk, endosperm, which is the white centre (85.5%), and the tiny germ (2.5%).

Typical biscuit flour is milled to a yield or extraction of 70–75%. Wholemeal flour is of 100% extraction and wheat meal flours in between these extraction rates, normally around 84% extraction. The flour will contain moisture of about 13–15%.

The wheat flour is composed of carbohydrate (as starch), protein and fat, together with some fibre, ash and trace minerals and vitamins. The protein is mainly gluten, composed of gliadin and glutenin. The percentage of protein determines the flour strength. A dough made from strong flour with a high protein content is extensible and can be machined into a continuous sheet for crackers and hard biscuits. A weak flour with a low protein content produces a soft dough, which may be moulded or deposited on the baking band and when baked, gives a short texture.

2.2.2 Wheat Flours: Typical Specifications

Table 2.1

Property	Soft flour (%)	Medium flour (%)	Strong flour (%)
Protein	8.0	10.3	13.2
Wet gluten	25.0	26.0	31.0
Fat	0.0	1.0	2.4
Carbohydrate	80.0	76.3	66.9
Ash	0.3	0.5	0.5
Water absorption	53.0	58.0	60.0

2.2.2.1 Wheat Gluten

The formation of the gluten, its strength and elasticity are largely determined by the flour specification, recipe and the mixing and forming processes. Wheat flour contains proteins including gliadin and glutenin. In the presence of water, these proteins combine to form gluten. As the dough is mixed the protein molecules form long strands of gluten, which have strength and elasticity. The gluten forms an elastic web, which gives the dough strength and allows it to be machined into a thin sheet for crackers and hard sweet biscuits. Crackers are made with 'strong' flour, which has a high protein content, typically 10–12%.

The gluten web is also important in trapping air and gas bubbles formed by yeast fermentation and by leavening agents such as sodium bicarbonate (soda) or ammonium bicarbonate (vol). This leavening process, combined with the laminating of the dough, gives the characteristic open, flaky texture of crackers during baking.

Soft or short biscuits are generally made with low protein flour (7–9%). A low protein flour makes a dough with a much weaker gluten web. In addition, these doughs have higher fat contents. The fat coats the flour particles and this inhibits the hydration of the proteins and the formation of the gluten web. Shorter mixing times also result in less development of the gluten strands, and hence the biscuits have a short texture.

2.2.2.2 Starch

Starch is the main component of wheat flour. It represents almost all of the carbohydrate content and around 80% of the total energy content of wheat flour. Starch is a polysaccharide (many sugars) made up of glucose units linked together to form long chains. The principal starch molecules in wheat flour are amylose, which typically comprises 28% of the total amount of starch. Amylose molecules contribute to gel formation. Their linear chains of molecules line up together and are able to bond to make a viscous gel.

Starch is insoluble in water; however, the starch granules do absorb a limited amount of water in the dough and swell. Above temperatures of 60–70°C, the swelling is irreversible and gelatinisation begins. The gelatinisation may continue until the starch granules are fully swollen, but it is normal in baked products that only partial gelatinisation occurs. The gelatinisation of the starch contributes to the rigidity and texture of the biscuit.

As the starch gel is heated further, dextrinisation occurs. This contributes to the colouring of the biscuit.

In soft dough products, the high sugar and fat content of the dough inhibits starch gelatinisation. The presence of sugar delays the gelatinisation of the starch, which may be due to the competition for the water. The fat, composed of triglycerides and surfactants, also tends to inhibit gelatinisation. With high sugar and fat recipes, the dough has a low gel viscosity and strength, and produces short and soft biscuits and cookies.

2.2.3 Sugar

Common sugar (sucrose) is a carbohydrate derived from sugarcane or sugar beet. It is a disaccharide composed of two monosaccharides, a molecule of glucose joined to a molecule of fructose.

Sugar gives sweetness, but it is also important in developing the texture of the biscuit. Dissolved sugar tends to inhibit starch gelatinisation and gluten formation and creates a biscuit with a more tender texture.

Undissolved sugar crystals give a crunchy, crisp texture. Sugar crystals, which melt during baking, cool to a noncrystalline glass-like state giving a crispy, crunchy texture, particularly on sugar-topped biscuits.

Dry sucrose melts at 160–186°C. Biscuits with sugar toppings which are melted to a smooth, shiny surface require high-intensity flash heat at the end of the oven to fully melt the sugar.

Invert sugar syrup is a mixture of glucose and fructose. The sucrose is split into its component monosaccharides by hydrolysis. The sucrose in solution is heated with a small quantity of acid such as citric acid. After inversion the solution is neutralised by the addition of soda. The invert syrup is sweeter than sugar and it contributes to a moist, tender texture in the biscuit.

2.2.4 Leavening Agents

Yeast is normally used in the production of cream crackers. The yeast is most active at temperatures of 30–35°C during dough fermentation. At temperatures above 40°C, the yeast activity stops and it is therefore inactive during the baking process.

Sodium bicarbonate (soda): Soda is readily soluble and it reacts with acidulants in the dough in the presence of water, producing carbon dioxide and decomposing to salt and water.

The speed of the reaction may be controlled by the type of acidulant used. The leavening of the dough takes place during mixing and fermentation of the dough.

Ammonium bicarbonate (vol): This leavening agent decomposes completely when heated, producing carbon dioxide, ammonia and water. The reaction is rapid at around 60°C, and therefore the expansion of the dough takes place during the initial stages of baking.

2.2.5 Fats

Fats are a vitally important ingredient in achieving the texture, mouth feel and the bite of the biscuit. Crackers and hard biscuits have relatively low percentages of fats in the recipes, while soft cookies have high amounts of fat.

Recipes with high fat contents require little water for producing a cohesive dough and produce soft, short doughs. During mixing, the fat coats the flour particles and this inhibits hydration and interrupts the formation of the gluten. Fats also tend to inhibit the leavening action of the carbon dioxide diffusion in the dough during baking, and this produces a softer, finer texture. Where both fat and sugar amounts in the recipe are high, they combine to make a soft, syrupy, chewy texture.

Typical blended vegetable dough fats are solid at ambient temperature and melt over a wide temperature range. Most fats used in biscuit making melt below blood temperature (36.9°C), and this avoids a waxy mouth feel. Fats are specified with a Solid Fat Index (SFI), which indicates the percentage of solid fat present in the total fat. A dough fat typically has an SFI of around 18 at 25°C and 12 at 30°C.

In baking a main concern with high fat recipes is the spread of soft cookies on the steel baking band, which is mainly due to the melting of the fat. This occurs very quickly as the dough pieces enter the oven and the temperature of the dough pieces increases above 35°C.

Note: For additional information on ingredients for biscuits, please refer to Appendix 1.

2.3 BAKING: THE DEVELOPMENT OF THE BISCUIT STRUCTURE AND TEXTURE

It will be seen from this brief consideration of biscuit ingredients and that there are complex chemical and physical changes taking place in the biscuit doughs and some of these are heat dependent. The changes that are temperature dependent mainly take place during fermentation and later during baking. These changes are also highly dependent on the moisture content of the dough and the humidity of the baking chamber.

As we have seen the water in the dough plays a vital role in achieving the biscuit texture and structure, it hydrates the protein allowing the gluten to form and develop and it hydrates the starch granules which swell and gelatinise.

The gluten can absorb a large amount of water. As the dough temperature rises, the gluten web swells and becomes strengthened and the structure of gas and air bubbles in the dough forms, causing an increase in volume of the dough pieces. The swelling of the proteins increases from 30°C to around 50°C. However, denaturation of the proteins takes place at temperatures over 50°C, when the long chains of molecules are broken. As more heat is applied, gluten coagulation occurs above 70°C. At this temperature some of the moisture is released from the gluten and contributes to the starch hydration and gelatinisation.

The air bubbles in the dough are saturated with water and these expand rapidly as the temperature increases. The increase in volume is 3% at 50°C, up to 50% at 95°C. This expansion creates a significant increase in volume of the dough piece during baking.

The hydrated starch molecules begin to gelatinise at temperatures of 50–60°C. In biscuits this process is partial as there is seldom enough water to fully gelatinise the starch. In short doughs with very little water, the starch gelatinisation is very limited. When the dough pieces have reached temperatures over 70°C, the structure is well formed and becomes stable although starch gelatinisation may continue until the dough reaches a temperature of 95°C.

In order for the biscuit to reach an optimum volume, it is essential that the surface of the dough piece is not dried too quickly, making it rigid and preventing the expansion of the dough piece. The dough piece surface must remain moist and flexible for as long as possible.

As the dough pieces, at ambient temperature, enter the oven, some moisture will condense on their surfaces. This not only keeps the surface of the dough pieces moist, but the condensation releases latent heat, which assists in raising the temperature of the dough. It is important to maintain a humid atmosphere in the first zone(s) of the oven and in some cases injecting steam into the baking chamber is also beneficial.

The physical and chemical changes noted above which form the texture and structure of the biscuit take place in the first half of the oven. They require not only temperature, but time as well. In some trials, it has been shown that there is a limit to the speed of the temperature increase, which if exceeded will result in a decline in the quality of the biscuit.

2.3.1 Moisture Removal

When the gluten and starch have been sufficiently hydrated and the structure of the biscuit is formed, the remaining free water must be evaporated. The water is evaporated from the surface of the dough pieces. This will occur principally at 100°C for pure water, but at higher temperatures (up to 130°C) when the water is held in solution, for example, in a sugar solution. At temperatures over 100°C, the application of heat will always result in moisture loss from the surface of the dough pieces, even in an oven atmosphere which is saturated with water vapour. This loss of moisture from the dough piece is dependent on the temperature, the method of heat transfer and the humidity of the oven (Table 2.2).

Table 2.2 Moisture loss

Ingredient	Cream cracker	Soda cracker	Snack cracker	Marie	Rotary moulded	Cookie
Flour	100.0	100.0	100.0	100.0	100.0	100.0
Sugar	1.8	0.0	6.0	21.4	25.0	50.8
Syrup	0.0	0.0	5.0	4.8	7.5	1.2
Fat	7.5	10.0	12.0	21.4	32.1	56.0
Leavening	0.9	0.8	2.0	1.3	0.5	1.0
Other	0.7	3.0	2.2	1.3	2.0	2.3
Water	31.0	32.0	20.0	20.0	9.0	19.1
Total	141.9	145.8	147.2	170.2	176.1	230.4
Total moisture	46.0	47.0	36.0	36.0	25.5	35.1
Dry weight	95.9	98.8	111.2	134.2	150.6	195.3
Final moisture (%)	2.5	2.5	2.5	3.0	3.0	3.0
Yield	98.4	101.3	114.0	138.4	155.2	201.3
Moisture to be removed per kilograms of biscuits baked	0.442	0.439	0.291	0.230	0.135	0.145

Approximate guide to the moisture removal for the main biscuit categories.

Cracker doughs have a large quantity of added water, typically around 15–22% of the total recipe. The final product will have a moisture content of around 2.5%, and this will require the removal and evaporation of 300–440 g of water for every kilogram of baked cracker.

The evaporation of this water requires heat transfer (the latent heat of evaporation), which is of 539 calories per gram of water.

As an example, the latent heat of evaporation required for baking 1200 kg/hour of snack cracker will be around 200,000 kcal/hour. This is clearly a significant energy requirement for cracker baking. It is, however, of much less significance for the baking of soft doughs and cookies. The equivalent energy requirement to provide the latent heat of evaporation for a soft dough biscuit would be around 87,000 kcal/hour.

2.3.2 Colour

There are several chemical and physical changes which contribute to the colouring of the biscuit surface. After the moisture has been mainly evaporated from the dough pieces, the temperature of the surface rises quickly and the colour will change from around 150°C. There are three processes which contribute to the browning of the biscuits.

Caramelisation is a nonenzymatic browning reaction, which is caused by the breakdown of sugars at high temperatures. The caramelisation of different sugars occurs at different temperatures: fructose at 110°C, glucose 160°C and sucrose at 160°C. Caramelisation results in both colour and flavour development.

A second browning process, dextrinisation, is the breaking down of starch molecules by heating. This produces pyrodextrins which are brown in colour and have a distinctive flavour.

Dextrinisation of the starch occurs at temperatures of 100–200°C.

The third browning process is known as the Maillard reaction. This is a complex chemistry in which many compounds are formed at high temperatures by the reaction of reducing sugars and amino acids. Since milk has a high content of proteins and amino acids, the Maillard reaction will also contribute to the colour of biscuits which have been brushed with milk before baking, giving a darker, rich brown colour.

These browning processes require high temperatures and occur when the biscuit surface is already dry. The colouring takes place during the final stage of the baking process.

2.4 SUMMARY

We have seen that there are many complex physical and chemical changes from the dough piece to the biscuit during baking. These changes are mainly temperature and time dependent and occur at different stages during baking. They may overlap and interact (Figure 2.4).

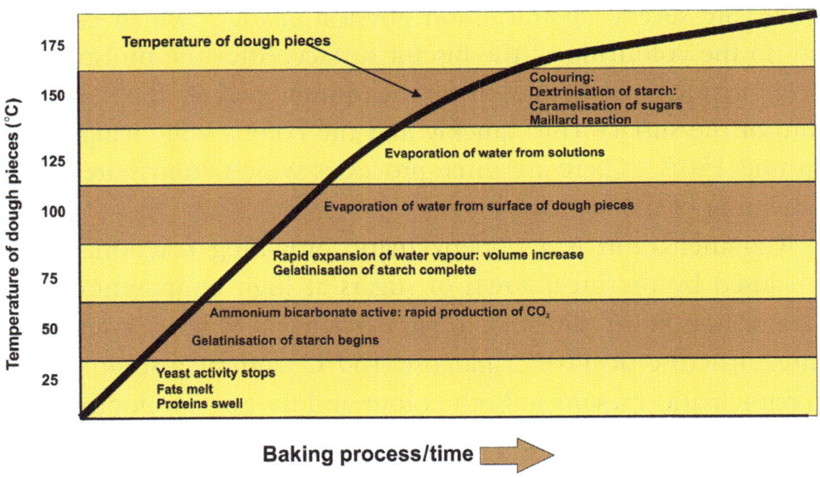

Figure 2.4 After W. Mowbray.

It is a useful simplification to say that the structure and texture of the biscuit is formed in the first half of the oven, that the moisture removal is in the middle and the colouring occurs at the final stage. The oven design should therefore provide a rapid heat transfer at the start of the bake and maintain a flexible, moist outer skin of the dough piece to allow expansion and lift. In the middle of the oven, the moisture should be removed efficiently from the dough pieces and then extracted from the baking chamber. In the final zone(s) the surface of the biscuit will be dry and it will colour. Good lateral heat transfer control is required to maintain an even colour and final moisture content of the baked biscuits.

The specification of the baking system should be based primarily on the product(s) to be made and their requirements in terms of structure, texture, density, bite, flavour and colour, a very bland even colour or contrasted background colour with highlights.

The characteristics of the biscuits will determine the type of heat transfer (radiation, conduction and convection) which is appropriate at each stage of the baking process. This will define the oven specification, the appropriate heat ratings and the zone lengths.

Figure 2.5 shows the heat ratings (heat energy input per square metre of oven band) for several multipurpose biscuit ovens.

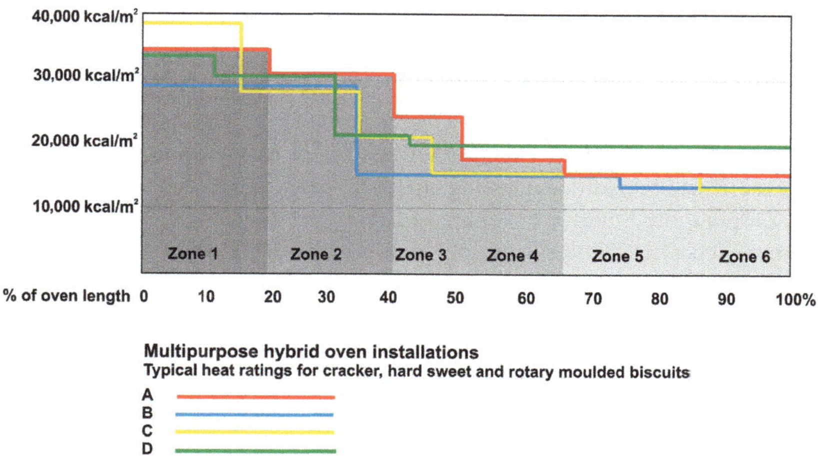

Figure 2.5

BIBLIOGRAPHY

Ayre, C., 2008. AB Enzymes targets improved biscuit baking. Decision News Media

Bender, A.E., 1990. Dictionary of Nutrition and Food Technology. Butterworths, Boston, MA.

Benedik, M., 2001. How does temperature affect yeast activity. <www.madsci.org>.

Buck, J.S., Walker, C.E., Sugar and Sucrose Ester Effects. University of Nebraska, Kansas State University.

CH Non Food, Wheat flour specifications. <www.ch-non-food.com>.

Dakota Baker's Yeast. Yeast fermentation in baked goods. <www.dakotayeast.com>.

Davidson, A., 2002. The Penguin Companion to Food. ISBN 0 14 2000163 5.

DTKFCPL, 2003. Examining the role of fats in bakery. <www.drtkfoods.com>.

Fineli, Wheat flour whole grain. <www.fineli.fi>.

Food-Info, Caramelisation. <http://www.food-info.net/uk/colour/caramel.htm>.

Food-Info, Caramelisation. Wageningen University. <www.food-info.net>.

Food-Info, Maillard reaction. Wageningen University. <www.food-info.net>.

Ghiasi, K., et al., 1982. Effects of flour components and dough ingredients on starch gelatinisation. Am. Assoc. Cereal Chem. 60 (1), 58–61.

Gurney, A., 2008. Bakery Fats and Oils. Leading Edge Bakery & Food Service Journal, 3197 Paterson Lakes, Victoria, Australia.

Johnson, J.M., et al., Starch. Oregon State University.

Lansbergen, G., 2002. Fats for Foods Consultants. Fat specifications. <www.fatsforfoods.com>.

Lowe, B., 2007–2009. The Boiling Point of Water and Solutions. StasoSphere.

Lowe, B., 2009. The Chemical and Physical Standpoint. StasoSphere.

Manley, D., 1996. Technology of Biscuits, Crackers and Cookies. Woodhead Publishing Ltd., ISBN 1 85573 280 7.

Manley, D., 2001. Biscuit, Cracker and Cookie Recipes for the Food Industry. Woodhead Publishing Ltd., ISBN 1 85573 543 1.

McCance, R.A., et al., 1945. The Chemical Composition of Wheat and Rye and of Flours Derived Therefrom. Department of Medicine, Cambridge.

Moodie, P., 2001. Traditional baking enzymes – proteases Enzyme Development Corporation. American Institute of Baking.

Penton Media Inc., 2008. A tropical touch, soya-based solutions. <www.baking-management.com>.

Robins, B.A., 1954. Formula, dough mixing method Review of Literature. Oregon State University.

Sofos Greek Produce Traders, 2009. Flour product List. <www.hktdc.com/manufacturers>.

Sumnu, G., et al., 1999. Effects of sugar, protein and water content European Food Research Technology. Springer-Verlag.

UGCS – Caltech, Dextrinisation. <www.ugcs.caltech.edu>.

CHAPTER 3

Biscuit Design and Output

Contents

The following notes introduce various factors of biscuit design and forming of the dough pieces which influence the baking operation.

3.1 CUTTER AND MOULDING ROLL LAYOUTS

The design of the cutting rolls and moulding rolls and dies for deposited cookies determines the pattern of dough pieces on the oven band. The cutting and moulding rolls are designed to give the maximum number of dough pieces per square metre of oven band. In the design of dies for cookies, allowance is made for the spread of the dough on the oven band during baking.

Normally the rolls are designed to provide a separation of about 8–10 mm between the edges of the biscuits on the oven band. The distance between dough pieces must also allow a sufficiently strong scrap dough lattice to be lifted without breakage after the cutter for cut biscuits.

Rectangular biscuits are baked with the short edge leading, which aids control during cooling, stacking and feeding to the packaging machines. Round biscuits may be 'nested' to gain the maximum loading on the oven band (Figures 3.1 and 3.2).

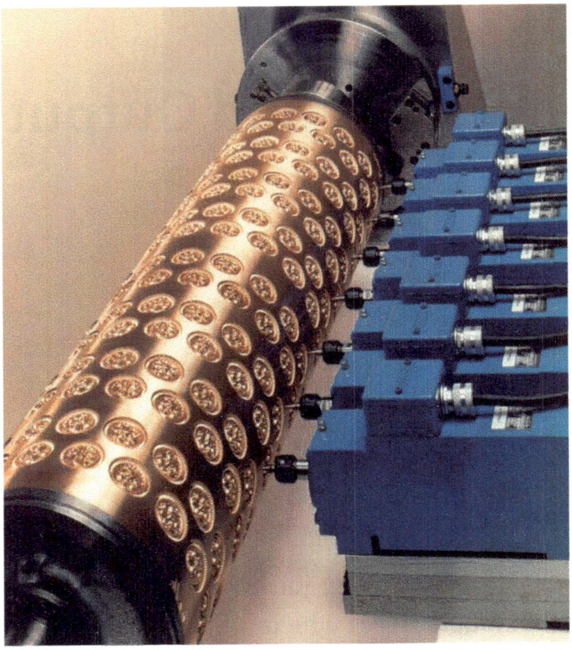

Figure 3.1 Rotary moulding roll engraving.

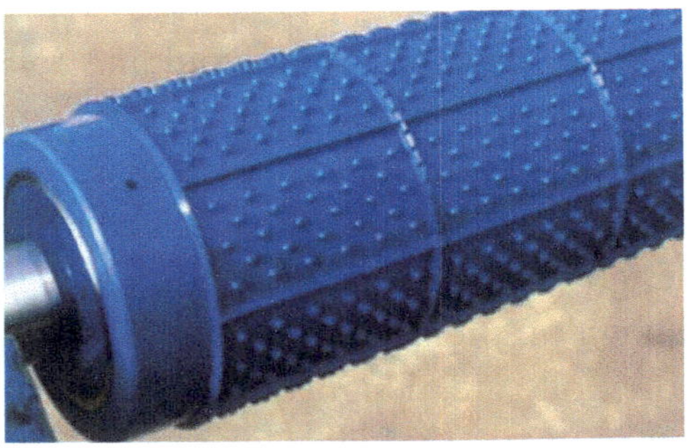

Figure 3.2 'Scrap-less' cutting roll with docker pins and perforated edges for the crackers.

3.2 SCRAP AND SCRAP-LESS DESIGNS

Hard sweet and cracker biscuits are produced from a continuous sheet of dough. The biscuit shape is cut, printed and perforated before being deposited (panned) onto the oven band.

Most products are cut into separate individual dough pieces before baking. The scrap dough around the dough pieces is recovered after the dough pieces are cut and returned to the sheeting machine at the start of the forming process.

Some products, notably soda crackers and some snack crackers, are baked in a continuous sheet. The dough sheet is perforated so that it can be easily and automatically broken after baking into the individual biscuits. When products are presented to the oven as a large sheet, there are several considerations:

1. The edges of the sheet at each side will pick up more heat from the edges of the band, which are not covered by the dough. The edge biscuits will therefore have more colour and this can be excessive. Oven band screens are used on some oven designs which deflect hot air away from the band edges. These are adjustable and will reduce the movement of hot air at the band edge and hence reduce the colour of the edge biscuits. In severe cases the biscuits may be baked with edge scrap dough which is removed after baking. Alternatively, steps may sometimes be taken to reduce the temperature of the band edges by forced cooling.
2. Cracker dough sheets usually shrink during baking and this can cause random breaks at the perforations on the edge of individual biscuits. These random breaks cause problems after baking at the automatic breakers as the biscuit sheets are presented irregularly. It is therefore worthwhile to reduce the size of the dough sheets and these may be cut through at approximately 1.0 m length by a large diameter cutting roll (approximately 320 mm in diameter). Alternatively, the crackers may be cut and baked in strips (Figures 3.3 and 3.4).

Most moulded and deposited cookies are formed and baked individually. Some extruded products, such as filled bars, may be baked in continuous 'ropes' and cut after baking. Layer cakes are baked in continuous sheets and slit and cut after baking (Figures 3.5 and 3.6).

Figure 3.3 Baker Perkins rotary cutter and scrap lift.

Figure 3.4 Crackers cut in strips.

Figure 3.5

Figure 3.6 Layer cake baked on a steel band.

3.3 DOCKER PINS

During baking, biscuits and crackers expand and lift to form a light open texture. During this process, rapid expansion of water vapour in the dough pieces occurs and the vapour needs to be released. This is accomplished by a series of holes in the biscuit design, called docker holes. The docker holes are placed in the design to release the vapour evenly and maintain a flat surface and even thickness of the biscuit. This accurate control of flatness and thickness is essential to the successful automatic packaging of the biscuits (Figures 3.7–3.9).

Figure 3.7 Biscuits with docker holes.

Figure 3.8

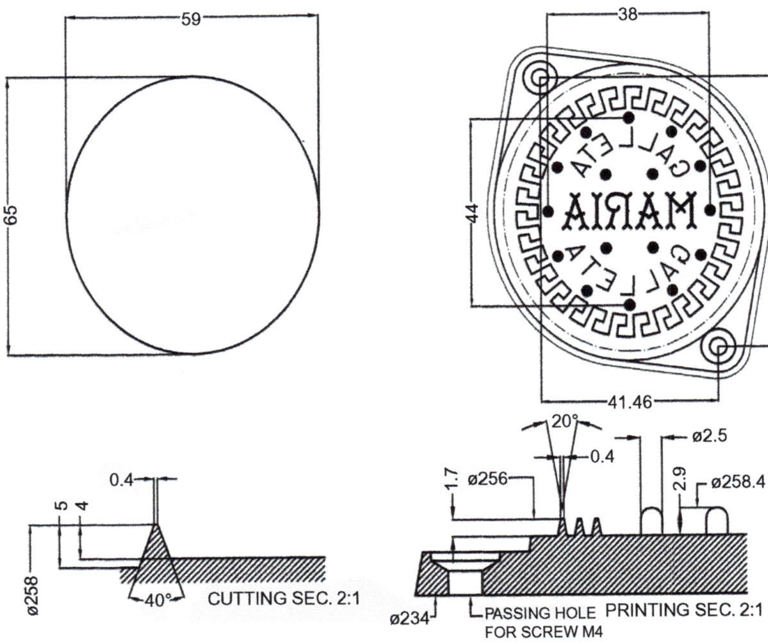

Figure 3.9 Cutter design for Maria showing the docker pin arrangement. Note the dough piece is cut as an oval shape to compensate for shrinkage during baking. Drawing and design by ErreBi Technology.

3.4 OVEN BAND LOADINGS

The band loading (weight of dough pieces on the oven band) will vary considerably depending on the biscuit design, band layout, biscuit weight and water content in the dough. The loading will influence the design of the oven band supports, drive and tracking system.

Biscuit	Biscuit		Oven band loading	
	Dimensions (mm)	Weight (g)	Dough (kg/m^2)	Biscuits (kg/m^2)
'Ritz' type cracker	48 diameter	3.0	1.40	1.04
Vegetable crackers	48 × 48	3.75	1.50	1.17
Soda crackers	91 × 44 (pair)	6.25	2.13	1.48
Marie	66 diameter	8.3	2.00	1.67
Glucose/'Tiger'	58 × 37	5.2	1.92	1.70
Butter cookie (moulded)	46 × 29	6.0	3.23	2.86
Wire cut cookie	50 diameter	6.5	2.14	2.00
Choc chip cookies	55 diameter	15	4.40	4.10

Approximate figures based on typical recipes.

3.5 OVEN SIZE AND OUTPUT

Usually the oven is the critical item in determining the capacity of a complete biscuit line.

Other considerations are mixing capacity, forming machine speeds, cooling and packing capacity; these are usually specified to suit the oven capacity.

The output of biscuits from an oven is determined by the baking time and the oven size. To determine the output from an oven, we calculate the number of biscuits across the width of the oven band and multiply this by the number of biscuits contained in the length of the oven.

This gives the total number of biscuits contained on the oven band during baking. We divide this by the baking time in minutes and this gives the total number of biscuits which will be baked in 1 minute. The output is usually expressed as the number of biscuits baked in 1 minute or in kilograms of biscuits baked in 1 hour.

For example, we can calculate the output for a typical rectangular moulded biscuit based on the following data:

Oven size (band width)	1200 mm
Oven size (baking chamber length)	60.0 m
Biscuit size	57 mm × 35 mm
Biscuit weight	4.5 g
Baking time	3.8 minutes

3.6 OUTPUT CALCULATION

Biscuits across the oven band	27 (allow 43 mm pitch)
Biscuits in the length of the oven	923 (allow 65 mm pitch)
Total biscuits contained on the oven band	27 × 923 = 24,921
Biscuits baked per minute	24,921/3.8 = **6558**
Oven output in kg/hour	6558 × 60 × 4.5/1000 = **1770 kg/hour**

3.7 SUMMARY

1. The cutting and moulding rolls are designed to provide the optimum oven band loading.
2. Consideration is given to the spacing between dough pieces on the band, the spread of cookies during baking and the orientation for the cooling and packing.
3. Some products such as soda crackers and snack crackers can be baked in a continuous sheet. Care must be taken to avoid excessive edge colour and irregular breakage of the dough sheet in the oven.
4. Docker pins are used to create holes in the dough pieces to allow the escape of water vapour and to control the surface

form of the biscuit, giving a regular flat surface with even blisters on crackers and hard sweet biscuits.

5. Biscuit output from the production line is usually determined by the size of the oven. The output is calculated from the biscuit size and weight, oven loading, oven band width, length of the baking chamber and the baking time.

BIBLIOGRAPHY

Baker Perkins Group, 2015. <www.bakerperkinsgroup.com>.

Errebi Technology Srl, 2015. <www.errebi.net>.

Manley, D., 1996. Technology of Biscuits, Crackers and Cookies. Woodhead Publishing Ltd., ISBN 1 85573 280 7.

Manley, D., 2001. Biscuit, Cracker and Cookie Recipes for the Food Industry. Woodhead Publishing Ltd., ISBN 1 85573 543 1.

CHAPTER 4

Heat Transfer

Contents

There are three modes of heat transfer: radiation, conduction and convection. All three heat transfer modes occur during biscuit baking. However, they differ in importance and effect on the product and their application depends on the oven design.

4.1 RADIATION

All objects above a temperature of absolute zero radiate energy to their surroundings. This energy or radiation is emitted as electromagnetic waves which travel at the speed of light.

The waves may travel through a vacuum or other medium. When they impact an object (other than a perfect 'black body'), they are partially absorbed and partially reflected. Good emitters are also good absorbers of thermal radiation (Figure 4.1).

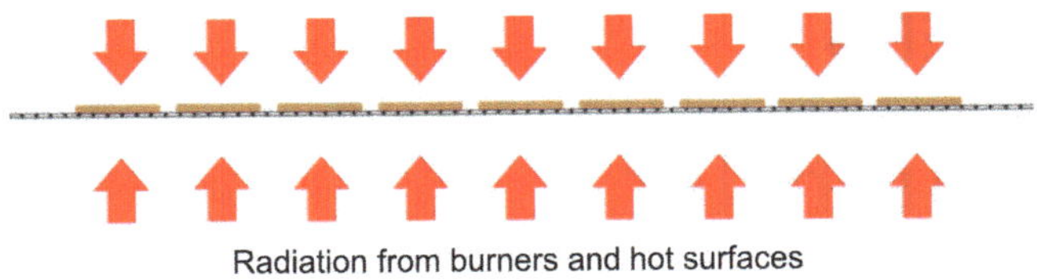

Figure 4.1

Electromagnetic waves vary in wavelength from infinitely short to infinitely long. They cover a spectrum from gamma rays (around 10^{-14} m) to radio waves (around 10^{2} m).

Visible light is a form of electromagnetic radiation with wavelengths of 0.40–0.71 µm (microns). Infrared radiation, which is important in baking, has wavelengths above visible light from 0.71 to 1.5 µm (near infrared) and from 1.5 to 4.0 µm (far infrared). Microwave baking uses longer wavelengths, normally around 12 cm (Figure 4.2).

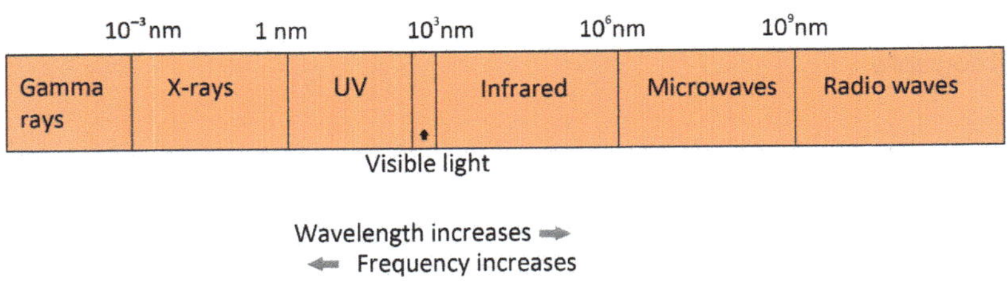

Figure 4.2

Thermal radiation is transferred from objects of a higher temperature to objects of a lower temperature. The electromagnetic radiation emitted by an object is directly related to its temperature. If the object is a perfect emitter (a black body), the amount of radiation given off is proportional to the fourth power of its temperature as measured in kelvin units (kelvin units= Centigrade temperature+ 273.15). For objects other than perfect black bodies, the power of radiation **P** is also proportional to the object's emissivity **e**. This relationship is described by the Stefan–Boltzmann Law:

$$P = e \times \text{sigma} \times T^{4}$$

where sigma = $5.67 \times 10^{-8} W/m^2 K^4$ and **T** is the temperature in kelvin.

Since the radiation is proportional to the temperature to the power of 4, a small increase in temperature will produce a very large increase in thermal radiation. The amount of radiation increases exponentially with a linear rise in temperature.

4.1.1 Wavelength

Most radiant energy emitted in a baking oven is within the infra-red spectrum and most of the energy emission is spread over a relatively narrow spectrum. The maximum emission is inversely proportional to its temperature. A higher temperature produces a shorter wavelength. The relationship is known as Wien's Law and is described by the following equation:

$$\text{Maximum wavelength} = C/T$$

where **C** is a constant equal to 2897 and **T** is the temperature in kelvin units.

If we consider a radiant oven with the hot tubes or ducts at 350°C, the wavelength will be $2897/(350 + 273.15) = 4.65\,\mu m$. This is a typical wavelength in an indirect radiant oven and it is in the infrared spectrum.

4.1.2 Distance

Finally the thermal energy radiated to an object is dependent on the object's distance from the source of the radiation. The intensity of the thermal radiation received is inversely proportional to the object's distance from radiating surface (inverse square law).

$$\text{Intensity} = I/d^2$$

where **I** is the intensity of radiation at the object's surface and **d** is the distance from the radiating surface. For example, if the distance is halved from 20 units to 10, the intensity will be multiplied by 4.

At 20 units distance: Intensity = $100/20^2 = 0.25$

At 10 units distance: Intensity = $100/10^2 = 1.00$

This is a very important factor in the design of an oven. The nearer the radiating surfaces are to the product, the more intense the thermal radiation will be and the more efficient the heat transfer.

4.1.3 Effect of Radiation on the Dough Pieces

When infrared energy impacts an object, such as a dough piece, its primary effect is to set molecules in vibration. The temperature of the dough piece is a measure of the average kinetic energy of its molecules, which increases due to the radiation from the oven surfaces in the baking chamber.

Objects such as dough pieces will absorb part of the infrared energy impacting on them and will reflect part. The energy absorbed penetrates the dough pieces and rapidly heats the centre of the dough pieces as well as the surface. This is an important attribute of radiation. This attribute is readily seen in microwave baking where the centre of a cookie is baked more quickly than the outer surface (Figures 4.3 and 4.4).

Figure 4.3 Radiant tubes in an indirect radiant oven.

Figure 4.4 Direct gas-fired burners.

4.2 CONDUCTION

Conduction is the transfer of thermal energy between neighbouring molecules within an object or between objects in contact with each other. It transfers energy from an area of higher temperature to an area of lower temperature and acts to equalise the temperature difference.

The Law of Heat Conduction is known as Fourier's law and this is shown in the following equation:

$$Q = -k\nabla T$$

Heat conduction, **Q**, is proportional to the thermal conductivity, **k**, multiplied by the negative temperature gradient (difference in the higher and lower temperatures), ∇ **T**.

Thermal conductivity **k** is the property of a material that indicates its ability to conduct heat.

The thermal conductivity predicts the power loss in watts through a piece of material. This is important in calculating the optimum insulation required for the oven.

4.2.1 Baking

The conduction of thermal energy to the dough pieces during baking is most relevant to baking on a solid steel band or a heavy mesh band such as a CB5 type. In these cases the dough pieces are deposited directly onto a hot oven band and the heat is conducted rapidly into the base of the dough pieces. Heavy mesh oven bands are normally preheated to a high temperature, 120–150°C.

Solid steel bands and CB5 bands have a relatively large heat mass and a large area in contact with the dough pieces. The conduction of heat to the base of the dough piece is optimised and contributes significantly to the total heat transfer. It is ideal for products such as soda crackers requiring high bottom heat to achieve good volume and an open texture. Conduction is also of major importance in the baking of soft cookies and cake on steel bands; the band conducts heat directly into the base of the cookies causing the fats to melt and the cookie dough to flow on the band to its final shape (Figures 4.5–4.9).

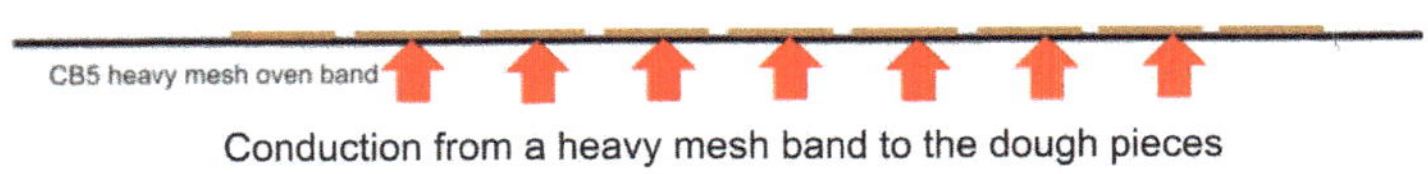

Figure 4.5

Figure 4.6 Cookie dough deposited on a steel band.

Figure 4.7 Cambridge Engineered Solutions: Compound Balanced Weave belt.

Figure 4.8 Dingson Food Machinery cracker oven with preheat burners located below the return band at the oven feed end.

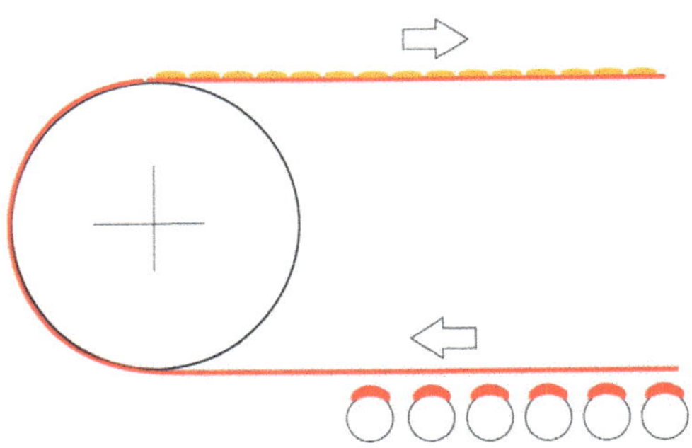

Figure 4.9 Diagram of preheat burners at the oven feed end.

4.2.2 Oven Insulation

The baking chamber must be effectively insulated to minimise the heat loss from the oven to the bakery. Heat is conducted from the walls of the baking chamber through the insulation to the outer covers. The amount of heat conducted depends on the material, quality and density of the insulation.

Materials such as mineral wool, normally used for oven insulation, have a low thermal conductivity. Mineral wool is manufactured from molten rock, stone, glass or slag, which is spun into fibres. The mineral wool is supplied loose or compacted into mattresses. The mineral wool typically has a thermal conductivity, **k**, measured in watts per metre kelvin of 0.06–0.10 W/(m K) at baking temperatures and is used at densities of at least 60 kg/m^3 (Figure 4.10).

Figure 4.10 Mineral wool insulation slabs for oven insulation.

4.3 CONVECTION

Convection is the movement that occurs in a fluid medium, for example, the air in the baking chamber. Hot air is less dense than colder air and will rise in the baking chamber. This is an important issue in the design of the temperature control systems since a change in temperature at the bottom of the oven will affect the temperature at the top.

Convection is described by Newton's Law of Cooling, which states that 'the rate of heat transfer is proportional to the differences in temperatures between the body and its surroundings.'

4.3.1 Convection Baking

The term convection is used in the biscuit industry to describe a heat transfer system employing jets of hot air blown directly on to the surface of the dough piece and the oven band from above and below. This hot air rapidly removes moisture from the surface of the dough pieces and increases the temperature of the surface of the dough pieces (Figure 4.11).

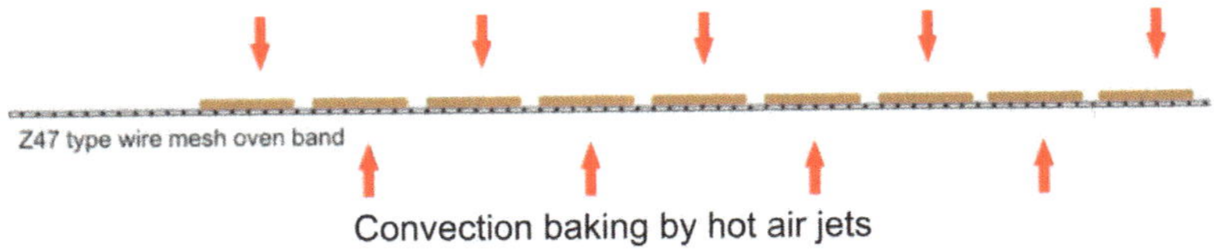

Figure 4.11

The convection system is very efficient in removing moisture during the middle part of the baking process. The hot, moving air constantly impinges on the surface of the dough pieces, rapidly evaporating moisture and removing it from the baking chamber through the extraction system. The convection baking system is usually controlled by altering the temperature of the air which is blown by the circulating fan into the baking chamber at a constant speed and volume.

It will be seen that this 'convection' system only affects the surface of the dough piece and primarily the top surface as the bottom of the dough piece is fully or partly shielded by the oven band. In the case of a steel band or heavy mesh band, the heat transfer to the base of the dough piece is solely by conduction. The effect of the impingement of the hot air jets on the surface of the dough pieces is to dry the surface rapidly, and this will form a dry, hard skin which will then rise in temperature and increase in colour. Since a large volume of hot air is being blown evenly across the dough pieces, the colour will tend to be even and will not show highlights and patterns as achieved by a radiant heat transfer system.

The rapid drying of the surface of the dough pieces forms a hard skin and prevents the expansion and 'lift' of the dough pieces so that this system is not used in the first part of the oven when baking crackers, most biscuits and cookies. In the first one-third of the oven length, convection should be at a minimum to prevent the skinning of the dough pieces and allow the texture, volume and shape of the products to develop.

4.4 SUMMARY

4.4.1 Radiation

Radiation is the most important method of heat transfer for biscuit baking. It occurs mainly by electromagnetic radiation of infrared wavelengths from direct gas burners, the hot surfaces of the baking chamber and tubes or ducts carrying hot gases from the burners. This radiant heat is penetrative and efficient and occurs without adverse side effects, such as the rapid drying or skinning of the surface of the dough pieces.

We have noted that the radiation (transfer of thermal energy) is dependent on several factors:

1. The nature and consequent emissivity of the radiating surface (how close it is to a 'black body').
2. The distance between the radiating surface and the dough pieces. The thermal energy transferred is inversely proportional to the square of the distance. A small decrease in the distance causes a large increase in heat transfer to the dough pieces.
3. The difference in temperature between the radiating surface and the dough pieces. A small increase in the temperature of the radiating surface causes a very large increase in the heat transfer. The heat transfer is proportional to the temperature of the radiating surface to the power of 4.
4. Radiation is the traditional method of heat transfer for baking biscuits and bread, as well as many other foodstuffs. It is the optimum heat transfer method for establishing quality texture, volume, shape and colour of biscuits. It is stable, penetrative and flexible, providing high heat inputs for crackers and a slow, 'gentle' heat transfer for soft doughs.

4.4.2 Conduction

1. Baking: Conduction transfers heat from the oven band directly to the base of the dough pieces. The heat transfer is dependent on the temperature and heat mass of the oven band and

the surface of the band in contact with the dough piece. With steel bands and heavy mesh bands this approximates to full contact. Ovens with band preheat can quickly transfer heat into the base of the dough pieces and achieve rapid development of the biscuit structure and texture; this is particularly valuable for cracker baking. Conduction is also important in baking soft doughs on steel bands.

2. Oven insulation: Heat from the baking chamber is conducted to the oven outer covers and contributes to the heat loss to the bakery. Insulation material with a low thermal conductivity is used to reduce this heat loss.

4.4.3 Convection

1. Convection baking uses hot air jets which impinge directly on the top of the dough pieces and the underside of the oven band. This system efficiently dries and colours the surface of the dough pieces. However, it produces a hard, dry skin on the dough pieces and will prevent good expansion and 'lift' of the product if used at the start of the baking process.
2. The convection of air in the baking chamber, where hotter air will rise, must be taken into account when controlling oven top and bottom temperatures independently.
3. Convection baking is favoured by oven manufacturers as it is relatively low cost to build and has a very simple control system. The burner power (total heat input to the baking chamber) is the main baking control in each zone. Most convection ovens have a fixed speed circulating fan and simple dampers to adjust the percentage of top to bottom airflow (Figure 4.12).

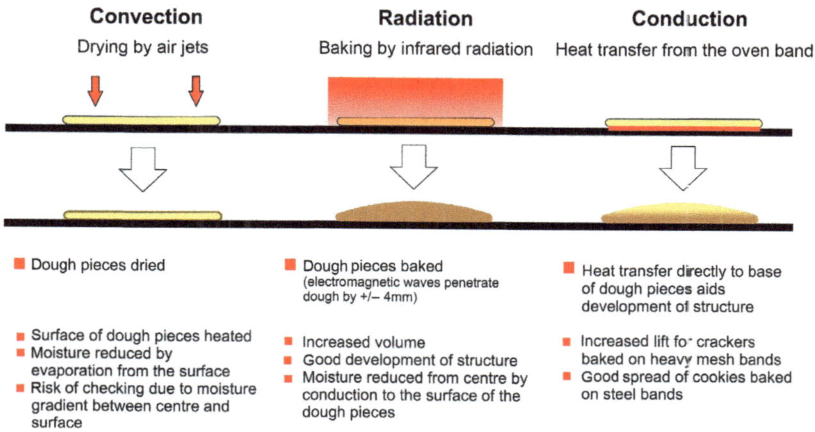

Figure 4.12 Summary of the effects of the modes of heat transfer.

BIBLIOGRAPHY

Department of Physics, 2010. University of Winnipeg. Heat transfer. <http://theory.uwinnipeg.ca>.

eFunda, 2015. Heat transfer. <www.efunda.com>.

HyperPhysics, 2010. Georgia University: Heat transfer. <http://hyperphysics.phy-astr.gsu.edu>.

McQueen Cairns, 2010. Hygrox. <www.mcqueen-cairns.com>.

Physicalgeography.net, 2010. The nature of radiation. <www.physicalgeography.net>.

Taftan Data, 2010. Thermodynamics. <www.taftan.com/thermodynamics>.

CHAPTER 5

Oven Designs

Contents

Biscuit baking ovens, generally known as tunnel ovens, have long conveyors which carry the dough pieces through a heated box section baking chamber. Oven lengths vary typically between 25 and 100 m long. The conveyor band material is a wire-mesh or a carbon steel sheet, which turns around large cylindrical drums at each end of the oven. The conveyor is driven by a variable speed drive at the oven end, which allows the operator to adjust the baking time.

5.1 HEAT TRANSFER METHODS

The baking chamber may be heated directly with gas burners or electric heaters or by an indirect system using heat exchangers. Direct heating systems use gas or electric energy; indirect systems may also use diesel oil fuel, as the products of combustion do not enter the baking chamber.

The temperature and humidity control is divided into zones along the length of the oven: usually each control zone is between 8 and 20 m long. This enables the temperature and humidity to be set and controlled throughout the baking process to optimise the conditions for the establishment of the biscuit structure, moisture content and colour as the dough pieces travel through the oven.

The control of the humidity in the baking chamber and the removal of moisture from the dough pieces are accomplished by an extraction system in each zone. This consists of ducts which draw air and moisture from the baking chamber through a fan and expel the air through vertical flues (chimneys) to atmosphere.

In some ovens, this wet air removed from the baking chamber can be diverted either to the flue or back into the baking chamber. This provides moving air within the baking chamber which can aid heat transfer and contribute to even baking conditions across the width of the oven. These systems are called 'turbulence' systems and are mainly used on ovens which have relatively still air in the baking chamber, for example, indirect radiant ovens and direct gas-fired ovens.

Ovens are designed to optimise the heat transfer to the dough pieces in different ways. We can group the basic designs into the following:

1. *Radiant heating*

   Direct gas-fired ovens, electric ovens and indirect radiant (cyclotherm) ovens (Figure 5.1).

Figure 5.1 Direct gas-fired/indirect radiant oven.

2. *Conduction heat transfer*

Ovens with preheated heavy mesh bands and steel bands (Figure 5.2).

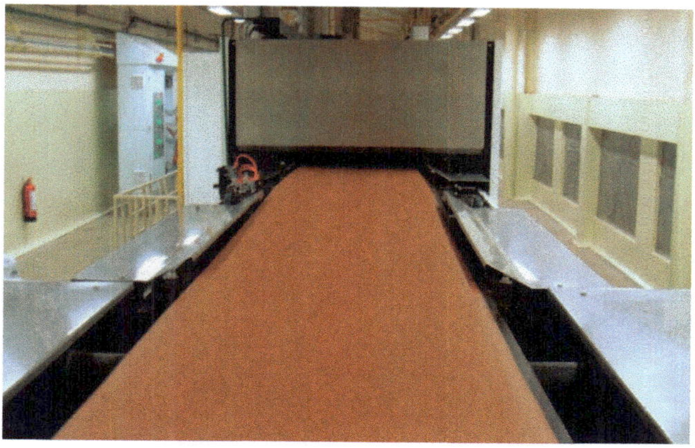

Figure 5.2 Cake baked on a steel band.

3. *'Convection' baking*

Direct and indirect convection ovens transfer the heat by blowing hot air jets onto the products. Also 're-circ' ovens which combine convection and radiant heat transfer (Figure 5.3).

Figure 5.3 Baker Perkins 'Jet Circ' oven.

5.1.1 Radiant Heating

5.1.1.1 Direct Gas-Fired Ovens

Direct gas-fired ovens are very widely used throughout the biscuit baking industry. They offer versatility to bake all types of biscuits, cookies and crackers (Figure 5.4).

Figure 5.4 Baker Pacific direct gas-fired oven.

The direct gas-fired oven has a simple baking chamber of box section with the oven band supported through the middle of the chamber. Above and below the band ribbon gas burners are located. A gas/air mixture is supplied to the burner tubes and this is ignited by a spark electrode and burns on a strip or ribbon across the width of the oven conveyor band.

The heat transfer in a direct gas-fired oven is primarily by radiation from the gas flames and from the oven top, base and walls of the baking chamber (Figure 5.5).

Most types of gas may be used including natural gas, town gas (manufactured from coal) and liquid petroleum gas.

The burners operate on a 'zero gas pressure' system and the pressure of the air supplied to the burners is controlled by motorised valves or variable speed blowers to increase or decrease the flame intensity and heat input. Various designs of corrugated stainless steel strips (or ribbons) are used to give a range of heat ratings. Woven wire-mesh (metal fibre) strips are also used for high infrared heat ratings.

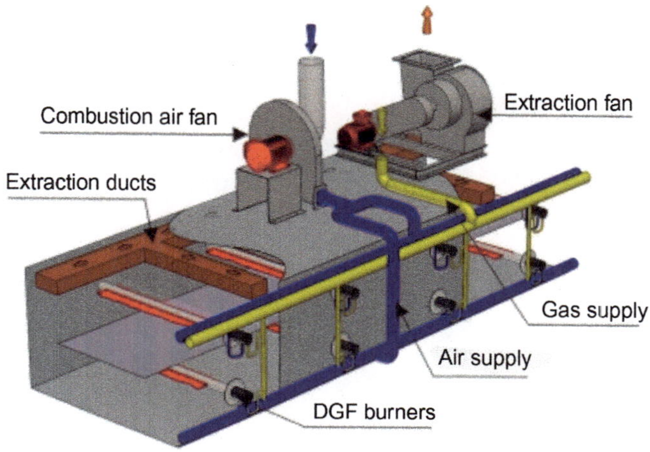

Figure 5.5 Direct gas-fired baking chamber.

This type of oven can achieve high heat inputs per square metre of band surface, up to 40,000 kcal/m^2 (46.5 kWh/m^2) may be used for cracker baking. Direct gas-fired ovens can successfully utilise any type of baking band: heavy mesh bands, open wire-mesh bands and steel bands. Direct gas-fired burners are used to supply band preheat where this is required for cracker baking (Figures 5.6 and 5.7).

The heat mass and stable radiant heat can be increased by installing heavy tiles at the top and bottom of the baking chamber. This arrangement is successfully used in the baking of water biscuits and crackers requiring very high heat inputs and stable radiant heating.

5.1.1.2 Summary

1. Direct gas-fired ovens provide radiant heat over a wide range of heat inputs.
2. Rapid and responsive control of heat input and accurate temperature control.
3. Suitable for all types of products: crackers, cookies, all types of biscuits.

Figure 5.6 Direct gas-fired metal fibre burners.

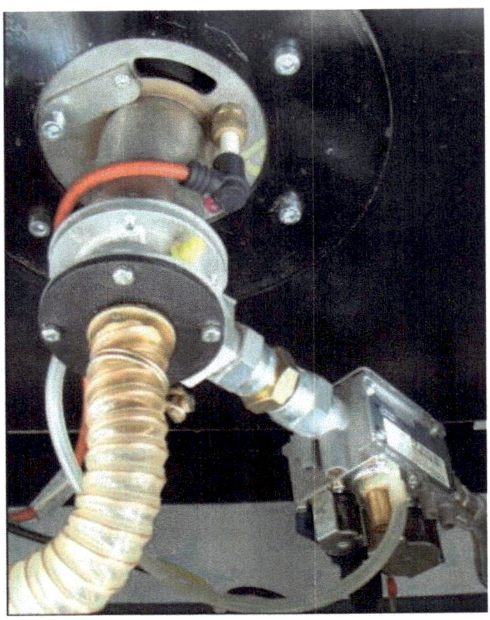

Figure 5.7 Burner head with gas/air mixer, solenoid valve, ignition and flame detection electrode.

5.1.1.3 Electric Ovens

Electric ovens are constructed in a similar way to direct gas-fired ovens, but use electric heating elements in place of the gas burners. These ovens have been widely used in the baking industry in some countries where industry had adequate electricity supply, but lacked gas, for example, China. However, most countries, including China, now use gas predominantly in the baking industry, which is invariably substantially cheaper than electricity.

Electrical heating has also been used in the first zones of ovens which required high heat inputs at the start of the baking process and where diesel oil with an indirect heating system was the preferred fuel for the main part of the oven (Figures 5.8 and 5.9).

5.1.1.4 Summary

1. Electrical energy is clean, does not require storage and is easily controlled.
2. Electric ovens can be simply controlled by thyristor units which vary the current in the elements and hence the heat input to the baking chamber.
3. Suitable for all types of crackers, cookies and biscuits.

Figure 5.8 Electric oven in China.

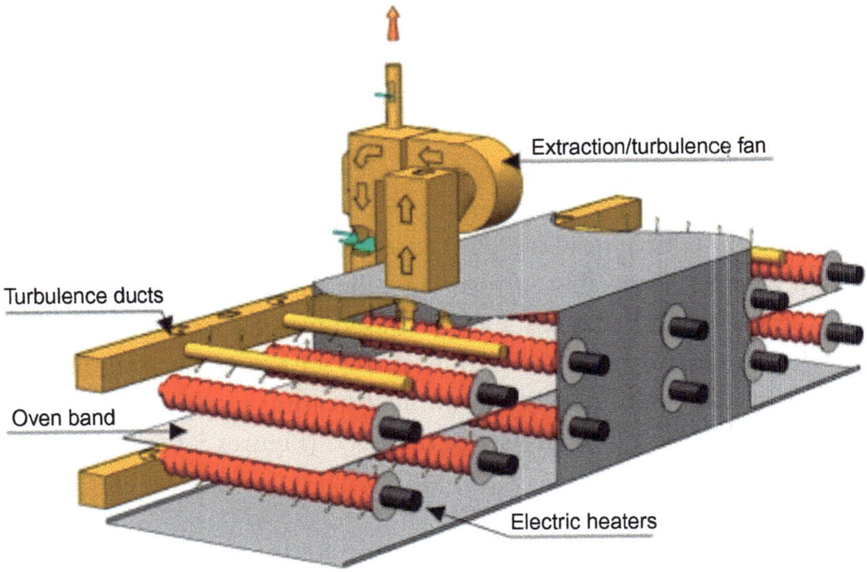

Figure 5.9 Electric oven baking chamber with turbulence system.

4. Provides a very dry atmosphere in the baking chamber and may require steam injection in the first zones for some products.
5. Very expensive to operate in most markets and not widely used now.

5.1.1.5 Indirect Radiant Ovens

Figure 5.10 Baker Pacific indirect radiant oven.

The indirect radiant ovens (also known as 'cyclotherm' ovens) are constructed in separate zones. Each zone is typically 10–20 m long and has a single burner, heat exchanger and circulation system for the hot gases from the burner.

Each zone has a burner firing into a burner tube. The hot burnt gases are drawn from the burner tube through ducts to rows of steel tubes, or ducts, at the top and bottom of the baking chamber. These radiant tubes, or ducts, run the whole length of the zone. The hot gases travel through the tubes or ducts which then radiate heat to the products from above and below. At the end of the zone, the hot gases are collected in a return duct through which they travel back to the circulating fan and from there to the burner tube to be recirculated. It is essentially a closed, circulating system with a single burner, circulating fan and radiant tubes to heat the products from above and below (Figure 5.11).

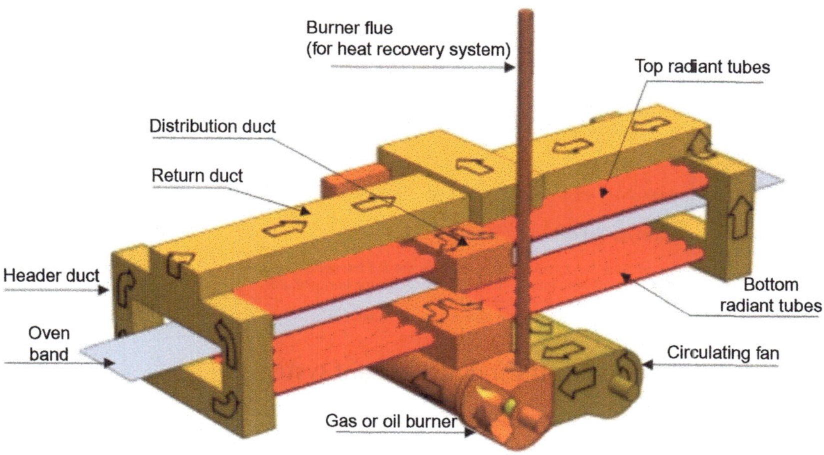

Figure 5.11 Indirect fired radiant heating system.

A flue with natural convection is used to balance the pressure in the system resulting from the ingress of combustion air at the burner. The continuous recirculation of the hot gases ensures a good efficiency. Fresh air is only drawn into the system at the burner for combustion and this is balanced by the natural extraction through the burner flue.

Since the products of combustion do not enter the baking chamber, the burner may use oil or gas. This system is commonly used where oil is the most economic fuel, for example, in India.

The indirect radiant baking system bakes by radiation with a high heat mass providing stable baking conditions. It is versatile, capable of baking all types of biscuit, cookies and some crackers. High-rate crackers require a first zone of direct heating. The system is favoured by many bakers for producing a high quality of biscuit structure, texture and colour. It is an ideal system for achieving colour contrasts on rotary moulded and cracker products (Figure 5.12).

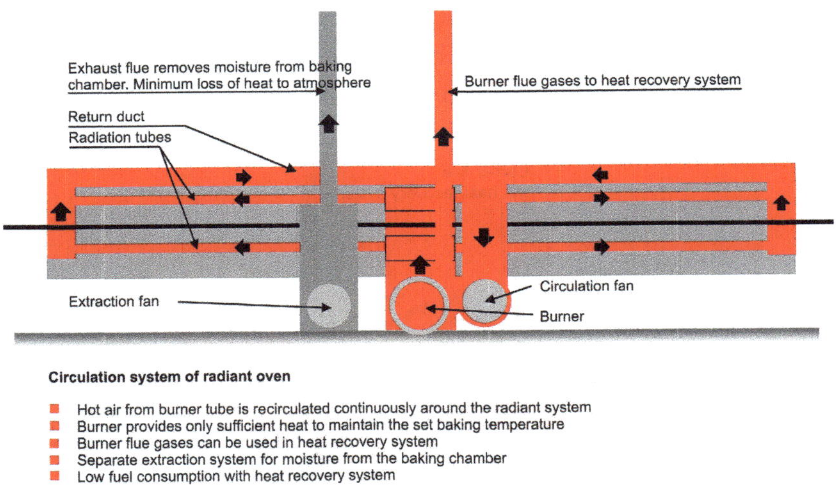

Figure 5.12 Summary of the indirect radiant system.

5.1.1.6 Summary

1. Indirect radiant ovens provide a very stable, radiant heat, preferred by many bakers.
2. The 'closed' circulation system retains and recirculates the hot gases from the burner and contributes to good fuel efficiency.
3. Suitable for all types of products, except for the first zone(s) of ovens for baking high-rate crackers (eg, soda crackers) which require direct heating.
4. Suitable for all types of fuel including diesel oil.

5.1.2 Conduction Heat Transfer

As noted above, steel baking bands and heavy mesh bands conduct heat rapidly into the base of the dough pieces. These types of band can be used in any of the oven designs, direct gas-fired ovens, indirect radiant ovens and convection ovens.

Steel bands are made of carbon steel, usually 1.2mm thick. They are principally used for the baking of cookies with high sugar and fat contents, which flow on the oven band in the first part of the oven. Traditionally steel bands are also used for the baking of 'Marie' biscuits (Figure 5.13).

Figure 5.13 Cookies deposited directly onto a steel band.

Heavy mesh baking bands, mainly type CB5, are woven with a tight 'herring bone' pattern providing a solid, thick and heavy mesh. These bands are preheated to 120–150°C and they conduct heat immediately into the base of the dough pieces as soon as they are deposited onto the band. This is a major baking method, being used throughout the industry for the baking of soda crackers and saltines. These bands are also versatile and can be used for a wide range of crackers, hard sweet and rotary moulded products (Figures 5.14 and 5.15).

Figure 5.14 Baker Perkins direct gas-fired oven with preheat burners.

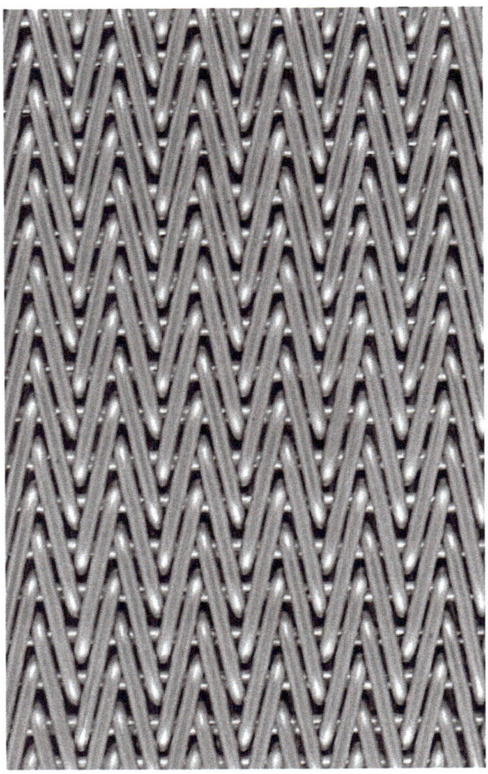

Figure 5.15 CB5 heavy mesh band from Yangzhou Jinrun Mesh Belt Manufacturing Co.

5.1.3 Convection Baking

'Convection' ovens are constructed in zones, each zone having a single burner and circulation fan. The fan blows the air around the burner tube or through the heat exchanger, where it is heated and then through ducts along the length of the zone. These ducts, located above and below the baking band, have slots or nozzles through which jets of hot air are blown onto the products and the oven band. Hot air from the baking chamber is drawn back to the fan to be recirculated through the system. Each zone has an extraction fan and flue to remove moisture from the baking chamber and this system will also extract some of the hot air from the heating system.

Typically, the discharge velocity of the air from the nozzles is a maximum of 20 m per second (approximately) at a maximum temperature of 310–320°C.

Convection ovens generally have fixed speed fans. The control of the baking process is therefore by temperature, by modulating the burner and by adjusting the proportion of convective air diverted to the top or bottom ducts. The airflow to the top and bottom of the oven is controlled by separate top and bottom manual or motorised dampers.

Temperature control is by thermocouple(s) detecting the temperature of the convective air.

The thermocouples are connected to temperature controllers in each zone which will automatically adjust the burners to increase/decrease the heat input.

5.1.3.1 Direct Convection Ovens

The hot gases from the burner are combined with the recirculated air from the baking chamber and blown through the ducts in the baking chamber and directly onto the products.

As the products of combustion are blown directly onto the products, diesel oil fuels are unsuitable for direct convection baking (Figures 5.16 and 5.17).

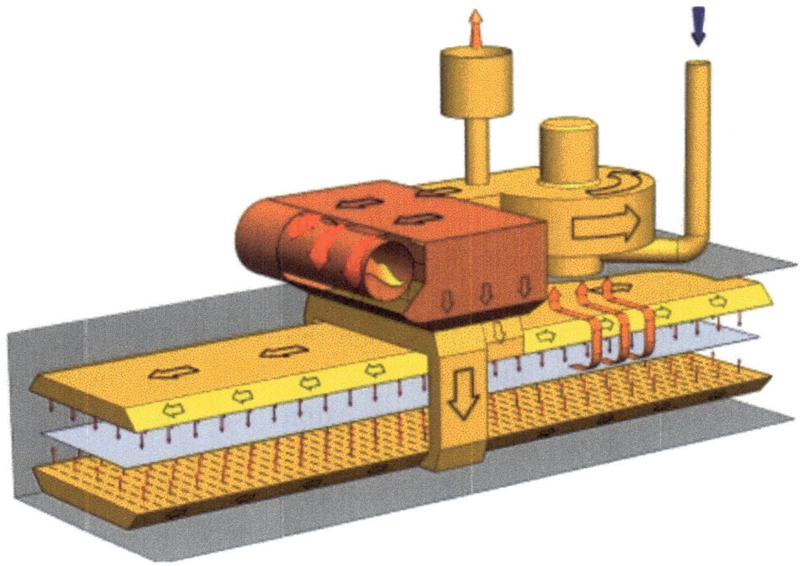

Figure 5.16 Direct convection baking system. *After Haas-Meincke.*

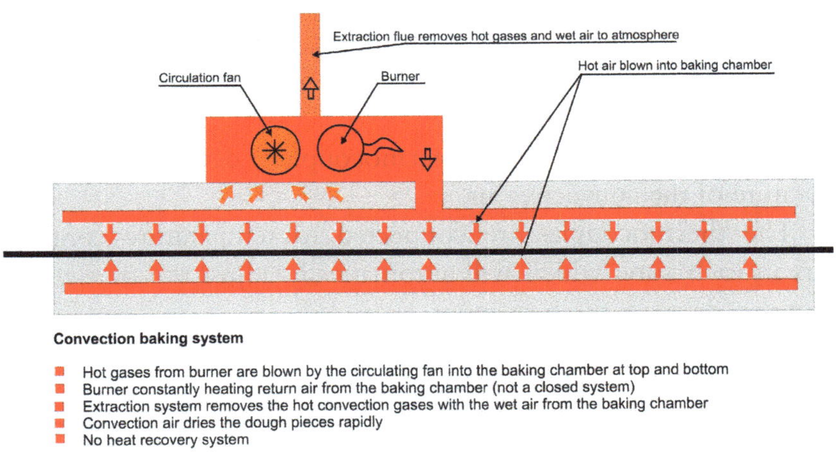

Figure 5.17 Summary of the direct convection baking system.

5.1.3.2 Indirect Convection Ovens

The burner fires into a burner tube connected to a multipass heat exchanger. The products of combustion are circulated within the heat exchanger and do not enter the baking chamber. Air is drawn from the baking chamber through the circulating fan and then passed through the heat exchanger, where it picks up heat, before being blow through the ducts in the baking chamber and onto the products. This system can utilise gas or diesel oil fuels (Figure 5.18).

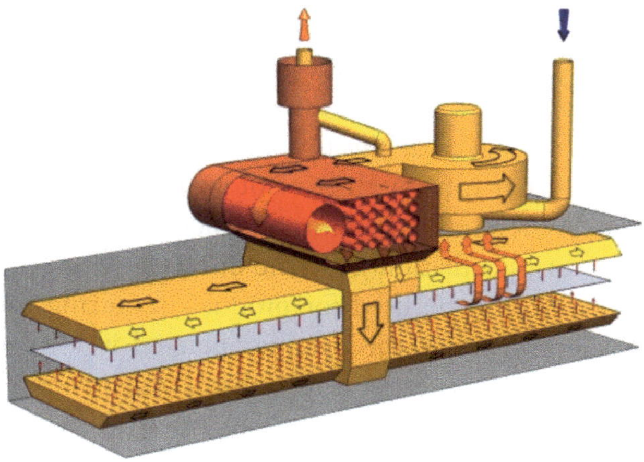

Figure 5.18 Indirect convection oven with heat exchanger. *After Haas-Meincke.*

5.1.3.3 Summary

1. Convection ovens bake by blowing hot air onto the product surface, so this is not a heat transfer system which optimises the development of biscuit volume and structure. However it is suitable and widely used for baking cookies, in particular Danish Butter Cookies.
2. Unsuitable for the first zone(s) of an oven baking crackers and hard sweet biscuits, as the convective air dries and skins the surface of the dough pieces preventing good lift and expansion.
3. Effective system for drying dough pieces to low moisture contents.
4. This rapid drying occurs at the product surface and may contribute to an excessive moisture gradient in the biscuit and consequent 'checking' or cracking of the packaged biscuits.

5. Provides even, bland colouring of the biscuits without contrasts.
6. Direct convection ovens are relatively low cost ovens to construct and control.

5.1.3.4 'Re-Circ' Ovens

These ovens were developed in the United States as a versatile cookie and biscuit oven. The system is basically a direct convection oven, but the volume of hot gases blown directly onto the products can be adjusted. When this is reduced, the hot gases circulate through the ducts at the top and bottom of the oven and return to the burner tube, with less hot air being blown onto the products. The system can therefore balance the heat transfer by convective air or by radiation from the ducts.

Re-circ ovens may be used for products requiring lower heat inputs. For the first zones, the oven is operated in a mainly radiant mode to avoid 'skinning' the dough piece. However, the heat transfer in this mode is low and the biscuit structure is relatively slow to form. In the convective mode the heat transfer is increased and the system operates as a direct convection oven (Figure 5.19).

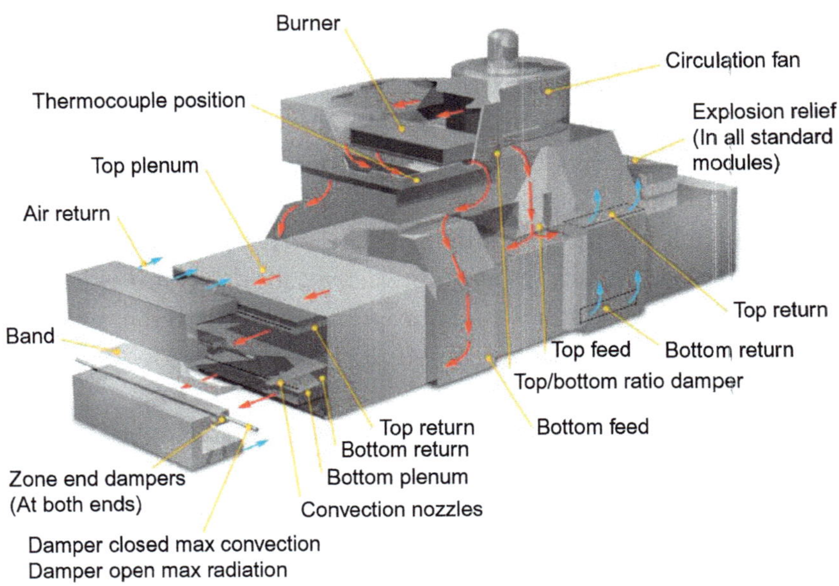

Figure 5.19 Baker Perkins re-circ oven.

5.1.3.5 Summary

1. Suitable for general baking of products requiring relatively low heat inputs.
2. Unsuitable for diesel oil fuels.

5.2 HYBRID OVENS

It has become common practice to combine different oven types into a 'hybrid' or 'combination' oven. This allows the baker to use different heat transfer modes at different stages of the baking process.

5.2.1 Direct Gas-Fired/Indirect Radiant Ovens

Products such as crackers and hard sweet biscuits require high heat inputs in the first part of the baking process to establish good structure and volume. This can only be provided effectively by a direct heating system and a direct gas-fired oven section is normally specified. This system also minimises the drying and skinning of the surface of the dough pieces, which would prevent the lift and expansion of the dough pieces when it is required.

The length of this direct fired section is usually one-third of the total length of the oven and the power input of the direct fired section is one-half of the total power input of the oven.

The indirect radiant oven section will contribute to the optimum development of texture and colour of a wide range of crackers and biscuits (Figure 5.20).

Figure 5.20 Combination oven: direct gas-fired/indirect radiant.

5.2.2 Direct Gas-Fired/Convection Ovens

This specification again uses the benefits of the direct gas-fired oven in the first one-third of the oven. The convection section will effectively remove moisture from the dough pieces and achieve a low and even moisture content for the final product. Colour will be even and bland, without contrasts, which is suitable for a range of hard sweet biscuits (Figure 5.21).

Figure 5.21 Dingson Food Machinery hybrid direct gas-fired/indirect convection oven.

5.2.3 Multisystem Ovens

Ovens utilising three systems, for example, direct gas-fired/indirect radiant/convection, have been installed. However, the theoretical benefits are outweighed by the complexity of set up and control and the relatively shorter sections of each system.

5.2.4 Dielectric Baking

Dielectric dryers have been manufactured by the Strayfield Company in the United Kingdom and used in the biscuit baking industry for almost 50 years. These are very effective in reducing moisture content and in achieving very even moisture contents across the width of the oven band. They are particularly effective on hard sweet biscuits which are prone to checking (cracking after baking due to stresses created by moisture gradient between the outer surface and the centre of the biscuit).

The Strayfield units are positioned immediately after the conventional oven and consist of a separate baking chamber, conveyor with synthetic band and drying system. The drying system comprises a transformer, rectifier, oscillator, circuit with variable inductance and electrodes above and below the product band.

The dielectric system subjects the products to direct heating from a radio frequency energy source. As the products pass between the electrodes, they are subject to a reversal of the polarity which causes the molecules of water to rub together and heat up. The heating rate increases as the frequency of the reversal of the charge increases. High frequencies are used, typically 27.12 MHz.

While the technology is effective, it is also relatively costly in terms of both capital cost and operating cost. It is therefore used mainly on high output lines where elimination of checking is a major and quantifiable benefit.

5.2.5 Microwave Baking

Microwave units have been successfully tested in conventional ovens. The addition of microwave units significantly speeds texture and volume development and aids moisture loss from the centre of the product. However, the cost of generating the microwaves and the safety issues involved have currently prevented the wider application of microwave baking in the industry.

BIBLIOGRAPHY

Baker Perkins Ltd, 2015. <www.bakerperkins.com>.

Coral, A., 2004. www.strayfield.co.uk Radio Frequency Heating and Post Baking. Strayfield Ltd.

Davidson, I., 1989. A baker's guide to modern biscuit ovens. Jpn. Biscuit Makers Assoc.

Dingson Food Machinery Ltd, 2015. <www.dsm-mc.com>.

Haas Meincke, 2014. <www.haas.com>.

CHAPTER 6

Oven Specifications: Hybrid Ovens

Contents

6.1 SPECIFICATIONS FOR OVENS: CRACKERS

A quality cracker has several important characteristics which are highly influenced by the baking process as follows:

- An open, light and flaky texture
- A surface colour which may have highlights (ie dark raised areas and lighter background areas)

- Toppings of cheese, herbs and spices which should have distinctive definition
- Low moisture content, around 1–2% (Figure 6.1).

Figure 6.1

How do we best achieve these characteristics and produce a biscuit of distinctive quality?

We will assume that the quality of ingredients, the formulation, mixing and forming processes have been carried out correctly. The baking is now critical to ensure that the dough pieces are baked to achieve the specific qualities of texture, moisture content and appearance required.

6.1.1 Development

The development of the light flaky texture occurs during the first 30–40% of the baking process. This change in structure involves the formation and expansion of gases and an increase in water vapour pressure which together give a dramatic increase in volume.

Simultaneously, softening of the fats, swelling and gelatinisation of starch and the denaturation of proteins occur. These changes cause the cracker structure to form.

The increase in volume must occur as the structure is forming but before the structure and particularly the outer surface becomes dry and rigid.

The dough pieces require a rapid high level of heat transfer, with maximum penetration. It is also necessary to delay the drying and setting of the top surface of the product as far as possible to allow maximum 'lift' and expansion. This requires that the baking chamber is humid and the surface of the dough piece remains moist. The humidity of the baking chamber may be assisted by the injection of steam at the feed end of the oven.

In order to achieve the correct conditions for good cracker development, it is necessary to apply maximum heat (of the order 40,000–50,000 k/cal/m^2/hour) to the bottom and by conduction to the center of the product immediately it enters the oven.

This is traditionally done in two ways. First (and this applies particularly to US practice of baking crackers such as soda crackers), by the use of heavy mesh bands which are preheated to temperature up to 150°C. These heavy bands conduct heat immediately to the bottom and then through the dough piece causing rapid expansion of the gases and water vapour and consequent lift of the product.

Alternatively (and we may use the example of traditional British practice in the baking of cream crackers), a lighter open mesh band is used and the bottom heat is applied by radiation from direct gas burners. Again high heat inputs of 40,000 k/cal/m^2 (46.5 kWh/m^2) of band are achieved.

The heat transfer from the burners is primarily through infrared radiation which penetrates the dough pieces.

6.1.2 Drying

Cracker biscuits also require considerable amounts of water to be removed. The dough pieces entering the oven may typically be at a moisture content of 15–25%, and this must be reduced to 1–2% in the finished product.

Drying is usually most efficiently carried out in convection zones. The moving air rapidly removes the moisture from the surface of the product. It also prevents a build-up of a static moisture barrier around the product which would inhibit further rapid moisture loss.

The main limitation on the speed of drying will be avoiding excessive colouring of the surface before the moisture content of the centre of the dough piece has been reduced adequately. If this moisture gradient is too high, stresses in the final product will cause checking (cracks which cause breakage of the biscuit after cooling).

We should however note that the velocity of convection air is normally up to 20 m/second, and this may be sufficient to disturb toppings on the product or even the product itself if these are small and light. In these cases a direct gas-fired (DGF) oven will be preferred.

6.1.3 Colouring

Colour changes occur when yellow and brown dextrins form and the sugars caramelize. This occurs at round 150–160°C. This may be achieved by a convection zone. However, we need to exercise caution since the use of forced convection systems can have two effects which may not be desirable.

Due to the nature of the heat transfer method (the impingement of air jets directly onto the product) the thin edges of the product may pick up colour very quickly causing 'ringing' (dark edges).

Also a highlighted colour may be required over the surface of the product. Since the convective air washes evenly over the whole surface of the product the resultant colour will tend to be even and bland. This is undesirable on a biscuit such as a cream cracker or a water biscuit which requires definite highlighting. This is achieved by predominant use of radiant heat in an indirect radiant or DGF baking section.

6.1.4 Final Moisture Content

The final moisture content of crackers is low (1–2%) and must be consistent. It is also important that the moisture gradient from the surface to the centre of the product is low.

In this respect the application of dielectric heating has been successful. The use of frequencies within the ranges of 13.56, 27.12 and 40.68 MHz give several inherent advantages in achieving low and even moisture contents in the final product. The heat is generated within the products in proportion to the moisture content, thus very even, consistent and low moisture contents can be achieved throughout the biscuit structure.

However, care must be exercised when baking products with inclusions or toppings such as cheese or fruit as these may present pockets of high moisture which may be burnt by dielectric heating.

Dielectric sections placed after a hybrid oven can often increase throughout and improve moisture control of the finished product. However, the dielectric units require the use of synthetic or plastic bands and are therefore constructed separately with their own band circuit.

6.1.5 Recommended Specification for a Cracker Oven

Specification of a combination oven for snack crackers (eg 'TUC' or 'Ritz' types) is shown in Table 6.1 and Figures 6.2–6.4.

Table 6.1 Recommended specification for a cracker oven

Process requirement	Heat transfer	Oven zone
Initial heating up	Radiant	DGF with preheat
Product development and formation of structure	Radiant	DGF
Drying	Convective	Convection or DGF with turbulence
Colouring	Convective or radiant	Convection or DGF with turbulence

Figure 6.2 Baker Perkins DGF cracker oven.

(a)

(b)

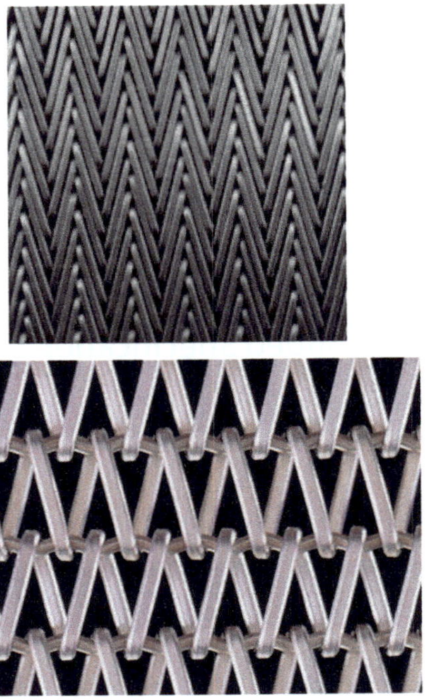

Figure 6.3 Cambridge Engineered Solutions. (a) compound balanced weave; (b) balanced weave.

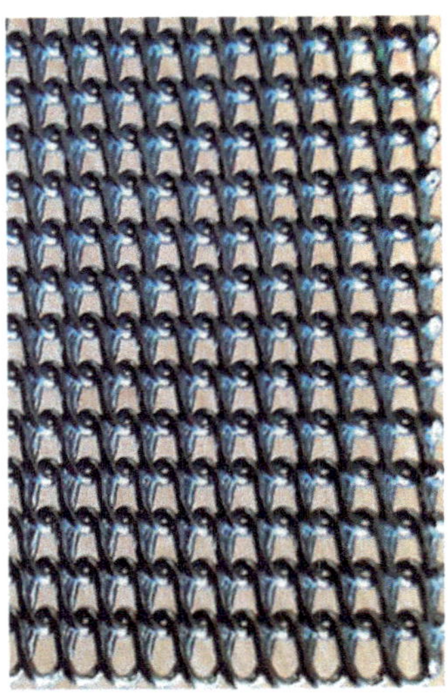

Figure 6.4 Steinhaus GmbH: F4012 (Z47 type).

6.2 RECOMMENDED OVEN SPECIFICATION FOR LIGHT CARRIER PRODUCTS (EG, CRISPBREADS, RUSKS AND BISCOTTES)

Figure 6.5

Dry and light carrier products represent a wide range but with a common process requirement.

They all require to be dried to even very low moisture levels. Reduction of moisture contents varies from 20% to 27% in the dough piece down to less than 1% in the finished product in some cases.

Good control for very even colour and establishment of correct structure is required.

However, the development of the structure does not demand the very stringent conditions required for crackers. Unlike soft dough biscuits, which often have a surface design and crackers which may have a highlighted appearance, many carrier products have a flat evenly coloured surface. These products are often formulated with wholemeal flours, rye and bran to enhance their healthy eating image. Many are therefore dark in colour.

The important characteristic is low and even moisture content. This can be achieved with several heat transfer methods provided the baking time is sufficiently long. However, the most efficient method is convection which achieves the best economies of fuel, space and the highest outputs.

Care must be taken in the final 30% of baking to prevent edge burning. A correct baking profile is therefore required. Minimal spacing of the product on the oven band or baking in continuous sheets will also reduce edge colouring and improve throughput. For this range of products, therefore convection ovens are the most efficient (Table 6.2; Figure 6.6).

Table 6.2 Recommended specifications for light carrier products

Process requirement	**Heat transfer method**	**Oven zone**
Initial heating	Radiation/conduction	DGF/heavy mesh band
Drying	Convective	Convection or DGF with turbulence
Colouring	Convective or radiant	Convection or DGF with turbulence

Figure 6.6 Baker Perkins "Jet Circ" convection oven.

6.3 RECOMMENDED OVEN SPECIFICATION FOR HARD SWEET BISCUITS (EG, MARIE)

Good quality hard sweet biscuits have the following characteristics:

- Good volume to give low density and a light product
- Smooth homogeneous texture
- Crispy bite
- Very even colour
- Low moisture content
- Very even moisture content to avoid checking (Figure 6.7).

Figure 6.7

An even gentle radiant heat is required to achieve good volume in the first third of the oven. The dough piece must not be 'skinned' or 'case hardened' and should increase substantially in volume. This gives a low-density light crispy bite.

Hard sweet doughs have considerable water content (around 20%), and this must be reduced to 2–3% in the baked product. It is essential to achieve even moisture content and avoid a moisture gradient from the centre to the outside of the biscuit, as this would lead to checking.

The baking time must be sufficiently long and gentle. This can be achieved by convection (low intensity) or by radiant heat from a DGF or indirect radiant oven.

The colour of hard sweet biscuits is generally very bland and even. This may be achieved by any of the main oven types.

For products with a serious checking problem, dielectric heating may be considered.

However, this is expensive in capital cost and running cost. Checking can be avoided by careful baking with adequate baking time on a conventional oven (Table 6.3; Figure 6.8).

Table 6.3 Recommended specifications for hard sweet biscuits

Process requirement	Heat transfer method	Oven zone
Development of structure and texture	Radiation	DGF
Drying	Convection	Convection or DGF with turbulence
Colouring	Convection or radiant	Convection or DGF with turbulence

Figure 6.8 Dingson food machinery direct gas-fired oven.

6.4 RECOMMENDED OVEN SPECIFICATION FOR SHORT DOUGH (ROTARY MOULDED) BISCUITS

Rotary moulded biscuits represent a wide variety of shapes, which need to be baked to give an attractive appearance (Figure 6.9).

Figure 6.9

- The short doughs have low moisture content
- Higher fat and sugar than hard doughs, less development of volume
- Generally baked on Z47 type wire-mesh oven bands
- Often baked with relatively short baking times, 3.8 minutes being typical
- Rotary moulded biscuits are often produced in large quantities.
- Many rotary moulded biscuits have a design requiring highlights, pale background and darker pattern. This is best achieved by radiant heat transfer (Table 6.4; Figure 6.10).

Table 6.4 Recommended specifications for short dough biscuits

Process requirements	Preferred heat transfer method	Oven zone specification
Formation of structure and texture	Radiant	DGF or indirect radiant
Drying	Radiant	DGF or indirect radiant with turbulence
Colouring	Radiant	DGF or indirect radiant with turbulence

Figure 6.10 Baker Pacific indirect radiant oven.

6.5 SPECIFICATIONS FOR OVENS: SOFT DOUGH COOKIES

Soft dough cookies, particularly those with large inclusions such as chocolate chips or nuts or centre fillings present quite different baking requirements.

The quality product must have the following characteristics:

- High degree of flavour, for example, fruit, nut or chocolate. These flavours must be preserved during the baking process.
- The flavour of the dough also, both its crust and its filling, must be enhanced.
- A distinctive eating quality which may require a soft centre and a crisp outer surface.
- An attractive surface appearance which may in some cases be gilded or sugar coated.
- A shape which may require a degree of flow on the oven band.

In general, soft dough products are much thicker and heavier than the crackers and with high fat content, and they may also have levels of moisture in the dough of around 10–15% (Figure 6.11).

Figure 6.11

Baking times are considerably longer, from 7–9 minutes for many deposited biscuits and up to 12–14 minutes for filled products.

The baking band also has an important influence on the heat transfer methods used. Soft dough cookies are baked on steel bands, and these act as a barrier to convective air. Heat will only be transferred to the base of the biscuit by conduction through the steel band.

During baking, the products which have high fat contents will flow on the steel band as the fats soften. This product flow takes place in the first part of the baking process. Subsequently, moisture levels must be reduced and finally the required colour and product appearance achieved.

Baking must be carried out slowly, gently and at relatively low temperatures in order to preserve the flavours and to allow good product flow on the band.

In summary, a heat transfer method which penetrates the thicker dough pieces is gentle and achieves an even balance of top and bottom heat application is required. In order to achieve these process requirements, radiant heat is preferred and a low baking temperature and relatively long baking time. The radiant heat may be applied by DGF or indirect radiant (cyclotherm) ovens. In order to achieve faster baking times, convective oven zones may be used in the final stages of baking, particularly when the even bland colouring produced is acceptable (Table 6.5).

Table 6.5 Recommended specifications for soft dough cookies

Process requirements	Preferred heat transfer method	Oven zone specification
Initial heating up, product flow and formation of structure	Radiant	DGF or indirect radiant
Drying	Radiant or convective	DGF or indirect radiant or convection
Colouring	Radiant or convective	DGF or indirect radiant or convection

6.5.1 Danish Butter Cookies

Danish butter cookies comprise a mix of rotary moulded and deposited products baked on a steel band. The most popular oven specification is a convection oven. The cookies are relatively small in size, spread very little on the steel band, doughs have low moisture content and the colour required is even. This can be achieved satisfactorily on any of the main oven types: indirect radiant, DGF or convection.

6.6 MODULAR OVEN DESIGN

By considering the process requirements of each product, we can therefore design an oven to enhance the particular characteristics of texture, eating quality, colour and appearance by the selective application of different methods of heat transfer.

We have proposed some optimum configurations of oven to produce certain quality products: snack crackers, dry carrier biscuits, hard sweet biscuits and soft dough cookies.

A modular oven range of ovens includes the following types of zone. Each type of module is compatible in size and construction with those of other types and can therefore be incorporated in the same biscuit oven (Table 6.6).

Table 6.6 Oven zone specifications to optimise heat transfer modes

Heat transfer method	Oven zone type
Radiation	Indirect radiant; direct gas fired
Convection	Convection ovens; turbulence systems
Conduction	We can use different band types to assist or reduce conduction to the base of the biscuit, eg, heavy mesh CB5, steel bands, wire-mesh
Direct product heating	Dielectric unit

6.7 CALCULATION OF OVEN ZONE LENGTHS

In order to calculate the best zone lengths for a particular oven specification, we need to consider:

1. The control we need to maintain the best baking profile for the oven. At certain stages of the baking process, we will need short zones to rapidly increase the baking temperature, at later stages the requirement may be a more steady, constant, stable

temperature requiring less heat input and therefore longer zones will be suitable.

2. Flexibility of control: We can specify the zone lengths and heat input for a specific biscuit. However, normally we will allow some flexibility to alter baking profiles and alter conditions to bake other products. More control zones give greater flexibility.
3. The heat input or rating from the burners in terms of kcal/m^2 or kWh/m^2 of oven band area. This is the heat intensity which can be transferred to the dough piece in each zone.
4. The construction of the oven modules: Manufacturers have standard module lengths, and zones are normally a multiple of these standard module lengths. For particular applications, special module lengths can be made.
5. Zone lengths normally vary between 8.0 and 18.0 m, though shorter or longer lengths may be made for particular circumstances.

6.8 EXAMPLE 1: DGF OVEN FOR BAKING CRACKERS (1.5 M × 100 M LONG)

Our baking profile requires a high heat input over the first 30% of the oven length to raise the baking temperature to 300°C and then to maintain a temperature of 280°C in the middle of the oven. The final colouring stage will require less heat input to maintain a baking temperature of 250°C.

The baker will determine the optimum baking profile for the product. Our task is now to translate the baking profile into a series of set temperatures and zone lengths. We will also consider maintaining flexibility in order to bake other products. For example, we can specify zones with the same set temperatures for one product but may require different settings for another product (Figure 6.12).

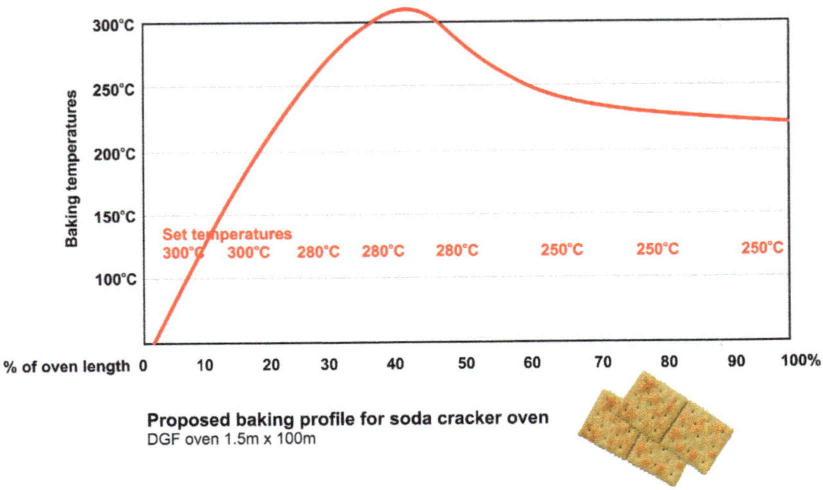

Figure 6.12

6.8.1 Heat Input

Baker Pacific oven modules are 2.9 m in length and contain 4 top burner positions and 4 bottom burner positions. Our DGF burners are rated at 20 kW and the high-rate burners at 30 kW. So in one module, we can input a total of 160 kWh with standard burners to a maximum of 240 kWh with high-rate burners (Figure 6.13).

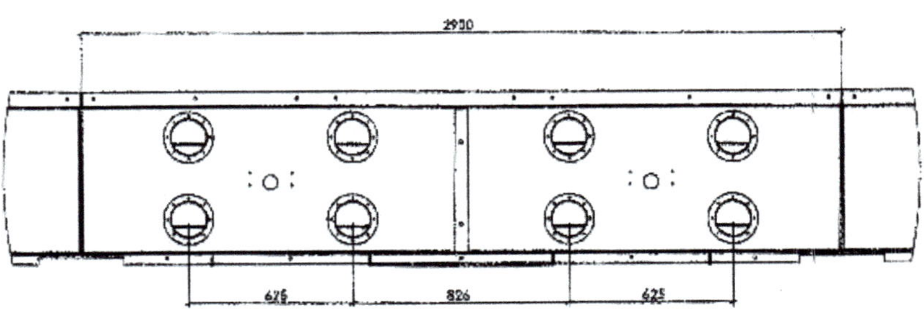

Figure 6.13

The area of oven band in the module is 1.5 m × 2.9 m = 4.35 m^2. The maximum heat rating that can be achieved is therefore from 36.78 kWh/m^2 with standard burners, up to 55.17 kW/m^2 with burners.

We can now divide the oven length into separate zones and specify the number and rating of the burners for each zone required to give the heat input to achieve our baking profile (Figure 6.14).

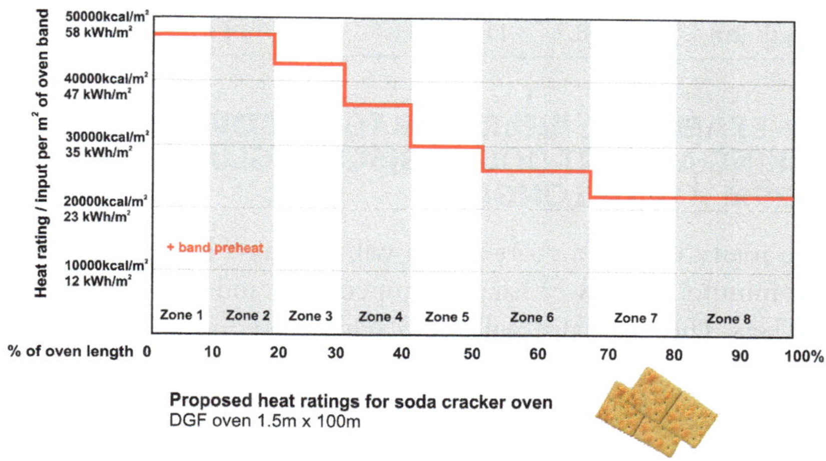

Figure 6.14

Our zone lengths will vary between 8.0 and 18.0 m, and they will be designed to achieve a rapid heat input in the first oven zones, to maintain a high temperature throughout the middle zones and then to reduce the heat input for the final zones. It should be noted that a considerable amount of heat is carried through the oven with the oven band and products, so the heat input required will reduce in the later oven zones, 5–8 (Figure 6.15). Our proposed oven zone configuration is shown above.

Figure 6.15 Baker Pacific DGF oven with 2.9 m modules and Eratec multifibre burners.

Zone lengths are

Zone	1	2	3	4	5	6	7	8
Length (m)	8.7	8.7	11.6	11.6	11.6	14.5	14.5	17.4

6.9 EXAMPLE 2: INDIRECT RADIANT OVEN FOR BAKING A SHORT DOUGH BISCUIT, 'GLUCOSE' (1.25 M × 100 M LONG)

The rotary moulded short dough will require a longer bake time (3.8 minute) and lower baking temperatures and heat inputs than crackers. The main aim will be to achieve a good biscuit texture, eating quality, flavour and colour. For this type of biscuit an indirect radiant oven is ideal (Figure 6.16).

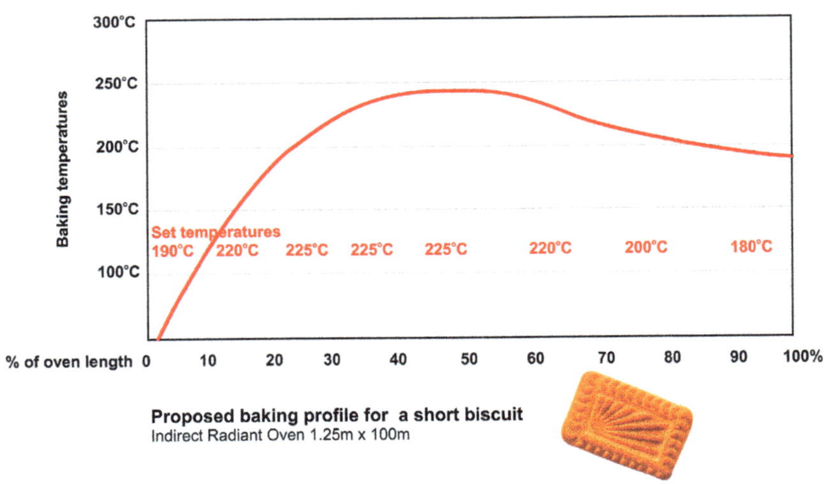

Figure 6.16

6.9.1 Heat Input

In an indirect radiant oven, there is one burner per zone. On a Baker Pacific oven, these are Weishaupt WGL-30 or Maxon OVENPAK 515SP burners with a maximum rating of 350–360 kW. The zones are made up from baking chamber modules of approximately 2.3 m long and heater modules 2.0 m long.

One burner supplies heat to each zone, so the heat rating is 350–360 kWh/area of oven band in the zone (Figure 6.17).

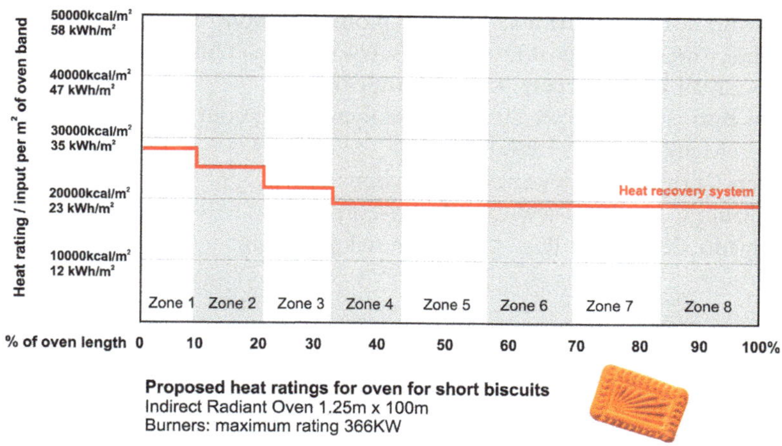

Figure 6.17

Zone lengths	1	2	3	4	5	6	7	8
Length (m)	8.9	11.2	11.2	11.2	13.5	13.5	13.5	15.8

The use of a heat recovery system enables the last two zones to be heated without burners (Figure 6.18).

Figure 6.18 Baker Pacific indirect radiant oven with heat recovery system.

BIBLIOGRAPHY

Baker Perkins Ltd, 2015. www.bakerperkins.com.
Cambridge Engineered Solutions, 2015. www.cambridge-es.com.
Digitron, 2015. Thermaflux. Thermal profile solutions. Data loggers for the direct measurement of heat flux. <www.digitron.co.uk>.
Eratec, 2014. MFB burners. <www.era-tec.fr>.
Flynn Burner Corp. USA, 2015. <www.flynnburner.com>.
Haas – Meincke, 2015. www.haas.com.
Maxon Corp, 2015. www.maxoncorp.com.
Weishaupt Corp, 2015. www.weishaupt-corp.com.
Wikipedia, 2010. Heat flux. <http://en.wikipedia.org>.

CHAPTER 7

Oven Construction: Direct Gas-Fired Ovens

Contents

7.1 DIRECT GAS-FIRED BAKING CHAMBER

The following are the main considerations in specifying and constructing a direct gas-fired (DGF) baking chamber:

- Baking chamber materials and dimensions

- Expansion joints
- Explosion relief
- Insulation
- Inspection doors
- Cleanout doors.

7.2 BAKING CHAMBER CONSTRUCTION

Baker Pacific chambers are constructed in 2.9m modules (Figures 7.1 and 7.2).

7.2.1 Materials

Baking chambers are normally constructed from mild steel, usually 2.5mm thick. This may be coated with heat-resistant aluminium paint applied only below the oven band to ensure no paint can flake and fall onto the product.

A special material may be used called 'Aludip', this is manufactured in the United Kingdom and available from Tata Steel.

Figure 7.1 Baker Pacific DGF baking chamber installation.

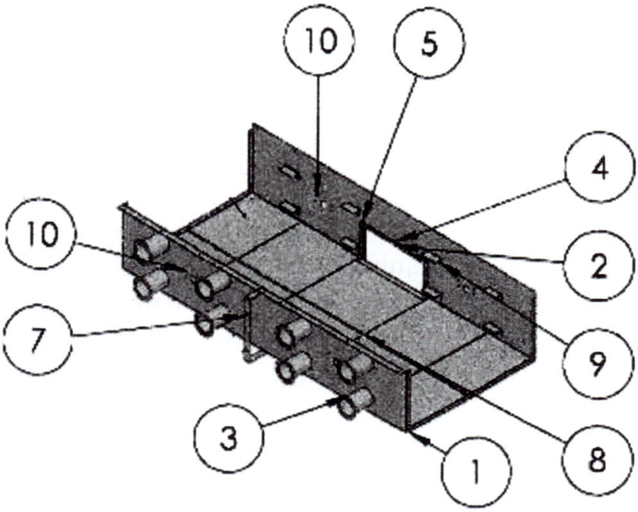

Figure 7.2 1 Baking chamber side, 2. Baking chamber side, 3. Burner box, 4. Cleanout door, 5. Cleanout door frame, 7. Cover support, 8. Base stiffener, 9. Burner supports, 10. Band support locations.

It is a mild steel sheet with an aluminium surface bonded to it at the mill.

The aluminium surface gives the material high corrosion resistance. Alternatives are special steels used in the building industry. These steels weather to a bronze colour and thereafter have good corrosion resistance.

For crackers and biscuits with salt sprinkled on top, it is usual to construct the first oven zone baking chamber in stainless steel.

7.2.2 Dimensions

The baking chamber should not exceed 550 mm in height to optimize the radiation intensity from the burners and the top and bottom of the chamber. Adequate clearance on each side of the oven band is required to allow tracking movement and the installation of tracking rollers.

Normally, this dimension is 150 mm on each side of the band.

7.2.3 Expansion Joints

As the baking chamber heats up and cools, it will expand and contract. In order to accommodate this movement, the baking chambers are allowed to slide on the base structure (Figures 7.3 and 7.4).

7.2.4 Insulation

The baking chamber requires adequate mineral wool insulation. This is supplied in slabs, 50, 100 and 150 mm thick and with 70 kg/m^3 density (standard) up to over 100 kg/m^3 density for particularly hot areas. The total thickness of the insulation should be 250 mm on top, 200 mm on the sides and 150 mm on the bottom of the bake chamber. The outer surface of the insulation uses mineral wool slabs which are covered by aluminium foil to protect them and to present a clean outer surface. Joins are covered by aluminium tape (Figures 7.5 and 7.6).

7.2.5 Oven Return Band Covers

On many DGF ovens, the return band is open and loses heat to the bakery. For cracker ovens with oven bands running at higher

Figure 7.3 Baking chamber support slide.

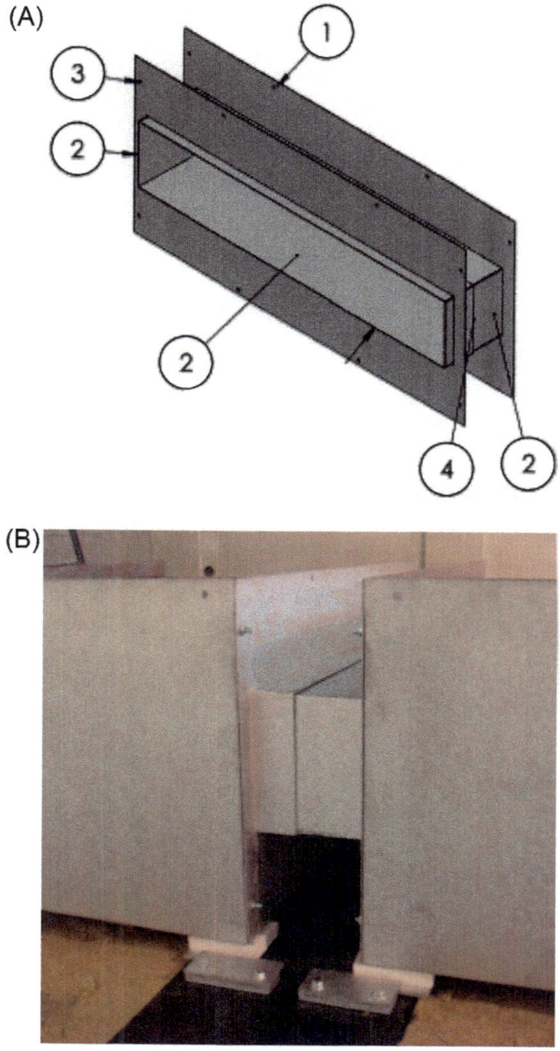

Figure 7.4 The joints between each zone have sliding spigots to take up the expansion. 1. Male backplate, 2. Male spigot, 3. Female backplate, 4. Female spigot.

temperatures, the return band should be covered with insulated panels. These may be easily lifted off for access to band support rollers (Figures 7.7 and 7.8).

Figure 7.5 Mineral wool insulation slabs.

Figure 7.6 Aluminium outer surface.

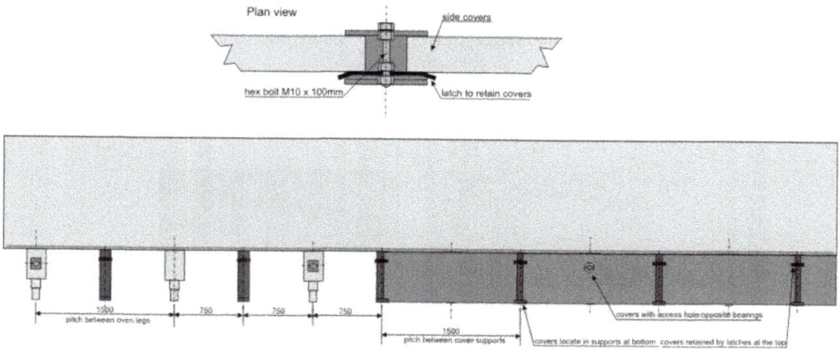

Figure 7.7 Construction of oven return band covers.

Figure 7.8 Oven return band covers.

7.2.6 Extraction

Moisture from the dough pieces may be extracted from the baking chamber in each zone. The wet air in the baking chamber is drawn into a series of cross ducts at the top of the baking chamber. This wet air then enters the extraction duct running along the top of the baking chamber to the extraction fan entry. The wet air is exhausted through a vertical flue (chimney) to atmosphere. The amount of air extracted in each zone is controlled by a damper, which may be manually adjusted or by a motorized valve controlled by PLC (Figure 7.9).

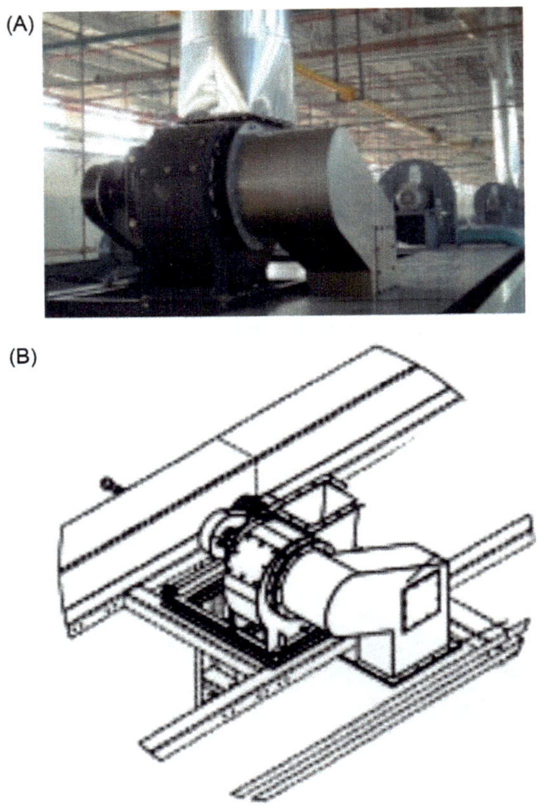

Figure 7.9 Halifax extraction fan mounted on top of DGF zone.

7.2.7 Fan Specification

Halifax 12P Multivane Forward Curved Fans (Figure 7.10)

Volume	34 m^3/minute
Pressure	70 mm
Outlet velocity	8.4 m/second
Impeller speed	1399 rpm

Figure 7.10 Direct gas-fired oven showing the extraction fans and combustion air fans.

7.2.8 Extraction: Oven End Hood

The oven band carrying baked products at the exit to the oven is 150–250°C and will lose heat to the bakery. In addition, hot air from the baking chamber is drawn out at the oven end.

In many cases, this area in the bakery is excessively hot. An enclosed hood with fan and chimney will assist in containing this heat and exhaust it through a flue to atmosphere (Figure 7.11).

Figure 7.11 Baker Pacific oven end hood with extraction fan.

7.2.9 Explosion Relief

In DGF ovens, there is a possibility of explosion or fire caused by ignition of unburned gas or by the product catching fire after a stoppage of the oven band and consequent overheating. In order to minimize the damage caused by explosion or fire, each baking chamber module has a relief panel in the top of the baking chamber. This has a thin plate, which represents the weakest part of the baking chamber and will blow out in the event of a rapid increase in pressure. The thin plate is retained by bars or chains so that it cannot be ejected and cause injury to nearby operators (Figures 7.12 and 7.13).

7.2.10 Inspection Doors

Each zone has an inspection door to enable the operator to check the condition of the product during the baking process. The door is provided with a lamp to view the product (Figure 7.14).

7.2.11 Cleanout Doors

Doors are provided on the non-control side of the baking chamber to provide access both for cleaning the inside of the chamber and for access to burner supports, oven band supports and tracking rollers (Figure 7.15).

Cleanout doors should provide access at least 600 mm × 300 mm, and the bottom of the door should be level with the base of the baking chamber to aid the sweeping out of debris. Cleanout doors should be provided in each module.

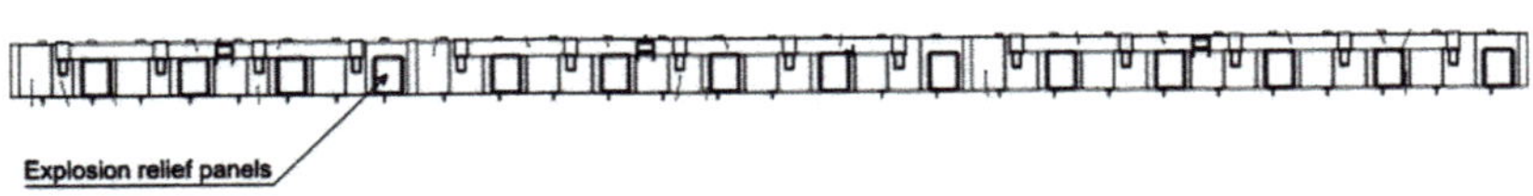

Figure 7.12 Plan of three oven zones showing the position of 14 explosion panels in the oven top.

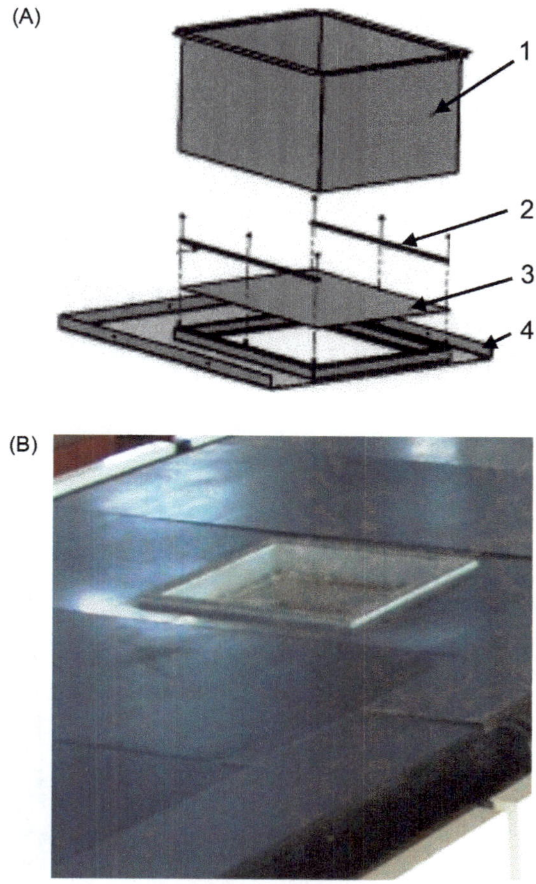

Figure 7.13 1. Explosion relief duct, 2. Retainers, 3. Explosion relief panel, 4. Top sheet of baking chamber.

Figure 7.14 Baker Pacific stainless steel inspection door.

Figure 7.15 Cleanout door on a DGF oven.

7.3 DGF OVEN: GAS BURNER SYSTEM

DGF oven burners use a 'zero pressure' gas system. The burners ignite and burn a gas/air mixture. The gas is supplied to the gas/air mixer unit at nominally zero pressure. The air is supplied at a positive controlled pressure. An increase in air pressure increases the flame and the heat input to the baking chamber.

The gas may be natural gas, town gas (manufactured from coal) or LPG (liquid petroleum gas).

7.3.1 Gas Train

Mains gas is supplied from a factory system to the oven where it is regulated by a 'gas train'.

This consists of the following equipment:

- Manual shutoff valve
- Two automatic shutoff valves for the safety of the system
- Gas filter
- Zero pressure gas governor
- Gas pressure high/low detection
- Gas pressure gauges (2) at gas inlet and outlet of gas train
- Gas valve tightness proving facility (to check for a leakage of gas)
- Main gas pipes and gas distribution system.

The gas is fed to header pipes running along each zone of the oven at the top and bottom. The headers are connected to each burner by flexible hoses and via a solenoid valve and gas/air mixer.

7.3.2 Combustion Air

In each zone the air is fed to the air header pipes from an air blower mounted on top of the oven. The air is drawn from the bakery and is filtered. The air pressure for the burner system is controlled by motorized valves which are regulated by the automatic temperature control system or by a variable speed inverter drive for the fan (Figures 7.16 and 7.17).

Figure 7.16 Combustion air fan mounted on top of a DGF oven.

Figure 7.17 Direct gas-fired burner installation with Eratec equipment showing: Gas header pipe (yellow); combustion air headers (blue); gas solenoid valves (also functioning as zero gas governors); gas/air mixers, flame managers.

7.3.3 Temperature Control System

In each oven zone the temperature in the baking chamber is detected above and below the oven band by thermocouples. There are two thermocouples above and two below the band in each zone. These are connected to a controller, which averages the temperatures to provide the actual baking temperature, either a single temperature or separate temperatures for the top and bottom of the oven. The control system will automatically increase or decrease the air pressure to the burners in order to increase/decrease the heat input to maintain the baking temperatures. The baking temperature should be maintained within ±1°C (Figure 7.18).

7.4 FLYNN BURNERS

DGF ovens use 'ribbon' burners installed above and below the oven band. A gas/air mixture is fed to the burner tube and burns on a special strip of corrugated or woven stainless steel. The gas may be natural gas, town gas (manufactured from coal) or LPG. In some cases an alternative design of strip may be required for town gas, which has a low calorific value.

The burner consists of a tube with a slot along part of its length into which are placed a number of formed stainless steel strips. This provides numerous small ports through which the gas/air mixture passes for combustion. This strip design ensures:

1. Continuous flame of uniform height
2. Good flame retention properties on high flame
3. Prevention of light back on low flame.

The burners installed in DGF ovens are generally of two sizes: 1″ diameter pipe with a capacity of 9.9 kcal/mm (11.5 W/mm) of strip and 2″ diameter with a capacity of 24.8 kcal/mm (28.9 W/mm).

The burners are installed in ports spaced approximately 625–825 mm apart. The burners are arranged with the strip at the side so that the flame is horizontal and is in the direction of the general airflow in the baking chamber from the feed to delivery end of the oven. The burners are ignited by spark electrodes (Figures 7.19–7.22).

Multilane distributor burners allow the flame to be adjusted in separate lanes to balance the baking temperature across the width of the oven band. The Flynn burner shown above is adjustable in three lanes. On DGF ovens, it is recommended that a minimum of 15% of the burners installed are adjustable.

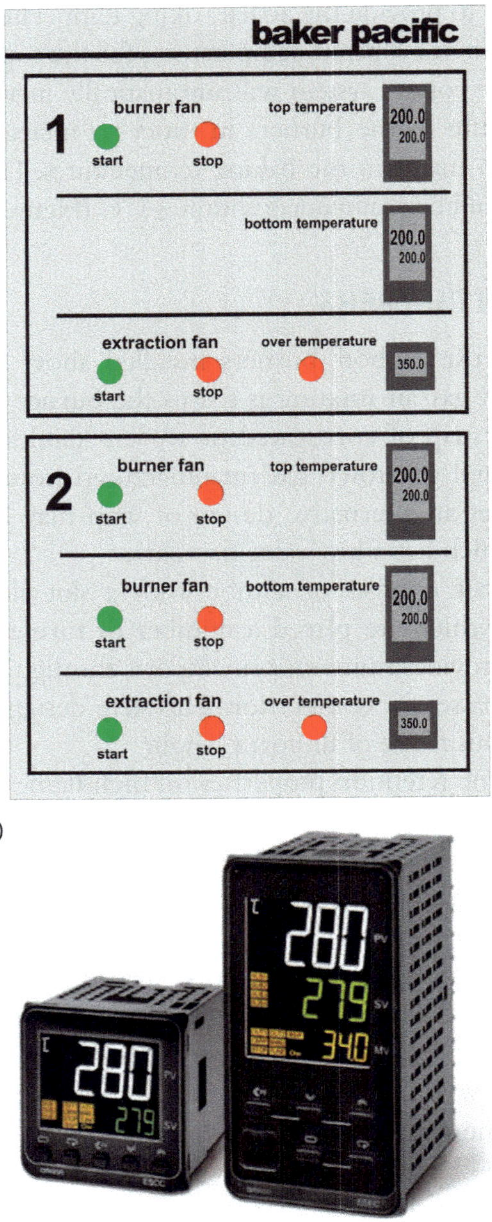

Figure 7.18 Baker Pacific zone control panel with Omron temperature controllers for two oven zones. The controllers show the set temperature and the actual temperature. There are separate controllers for the top and bottom temperatures and a pre-set controller to detect over temperature for safety.

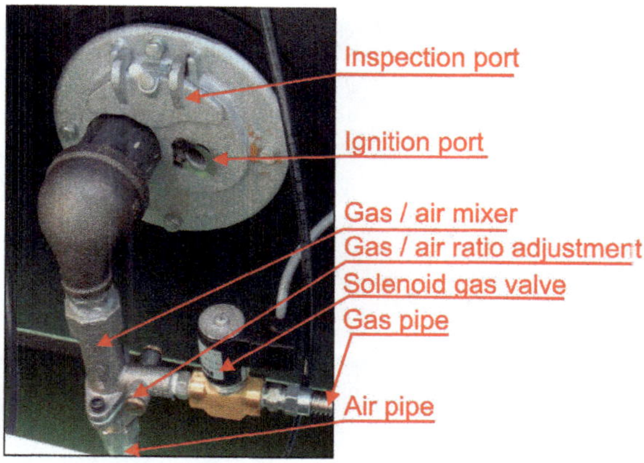

Figure 7.19 Flynn burner installation.

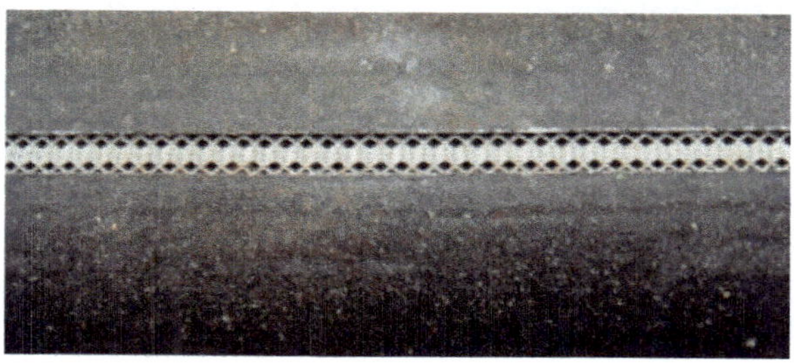

Figure 7.20 Burner strip.

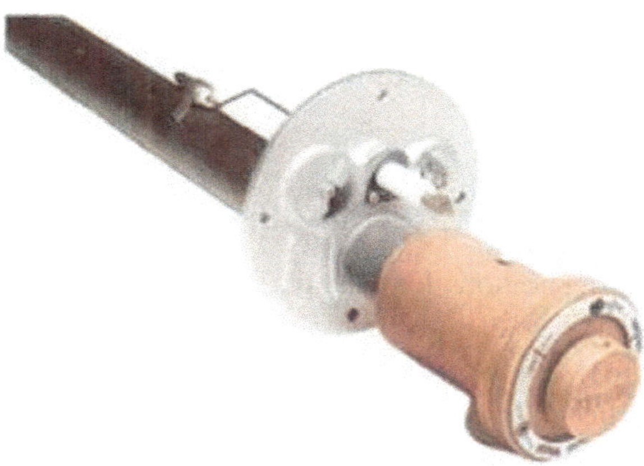

Figure 7.21 Flynn distributor burner head (three lanes).

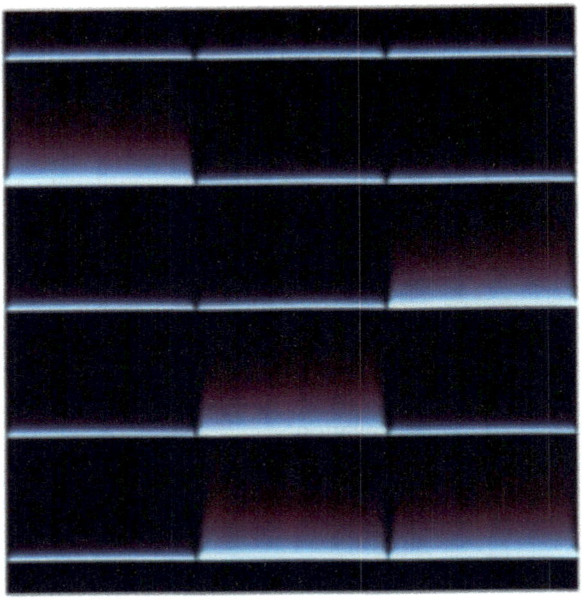

Figure 7.22 Flame adjustments in three lanes.

7.4.1 Flame Manager

The flame ignition is controlled at each burner by a flame manager/spark monitor system (Figure 7.23).

This solid state unit controls the electrical spark ignition and monitors the flame. The unit is equipped with on/off switch and flame-indicating lights. It can be used with most existing ignition electrodes.

7.5 ERATEC MFB (MULTIFIBRE) BURNERS

For high-rate applications, for example, cracker ovens, high-rate burners may be used. These use a 'knitted' mesh strip (Figures 7.24–7.26).

High-rate MFB burners may be used effectively as bottom burners in the first zones of the oven. The additional high radiant heat aids rapid lift and expansion of the crackers, resulting in

Figure 7.23 Flynn spark monitor system.

Figure 7.24 MFB burners in a DGF oven.

Figure 7.25 Gas/air mixer and Siemens solenoid valve.

Figure 7.26 Baker Pacific DGF oven with MFB burners.

a more open texture and faster baking. Eratec MFB burners have the following features:

- Direct heat transfer by radiation (without contact and air movement)
- High radiant power density 100–1000 kW/m^2
- Precise control and power modulation
- Energy consumption savings compared to conventional DGF burners (up to 20% saving depending on the oven and number of burners)
- Low pollution (up to 80% less CO and NOX)
- Safe for people and equipment (no burn back)
- Low maintenance

7.5.1 Burner Specification

- Cylindrical infrared burner in knitted FecralloyR Nit200S from Bekaert
- Length of IR surface: 1000–1500 mm
- Casing: tube stainless steel 304: 2–60.3 mm diameter
- Input power from 6 to 34 kW
- Working position: horizontal–up/down radiation adjustment
- 1-Electrode (ignition and detection)
- Input power adjustment by air pressure from 20 to 80 mbar
- Gas pressure 0 bar
- Available with multilane adjustment

7.6 CONTROL PANEL FASCIA DESIGNS

7.6.1 Main Control Panels

These panels, located at the end of the oven, will contain all the controls and displays for the conveyor drive, band cleaner, tracking and tension, burners off/on and purge cycle, together with safety alarms (Figures 7.27 and 7.28).

Figure 7.27 Zone panel and main oven control panels.

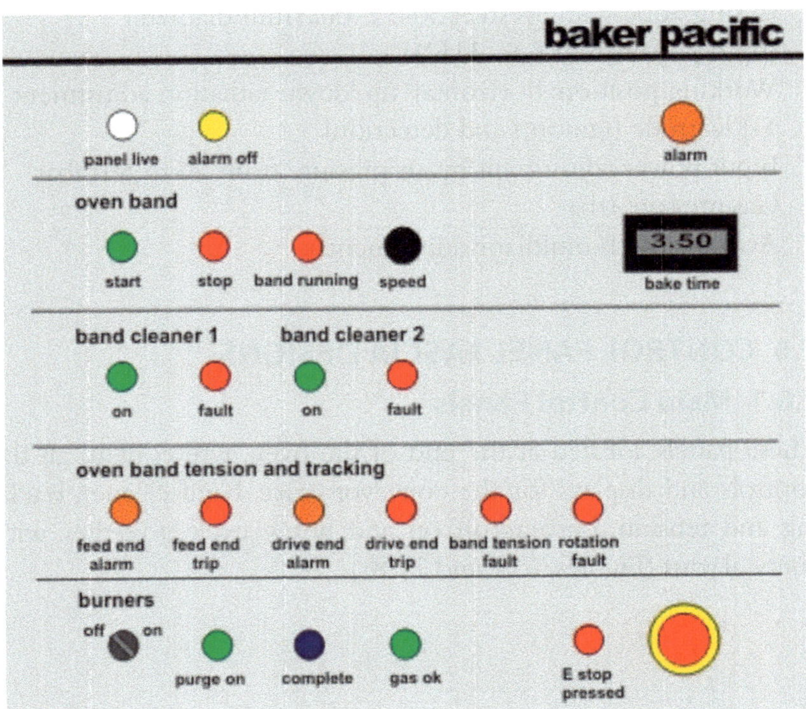

Figure 7.28 Main panel fascia drawing.

7.6.2 Zone Control Panels: DGF Oven

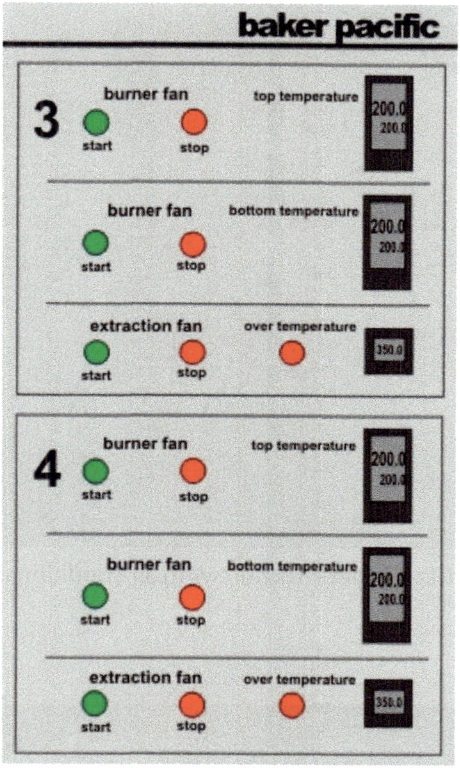

Figure 7.29 Zone panel for separate top and bottom burner control.

7.7 CONTROL PANEL CONSTRUCTION

The industry standard is Rittal, Friedhelm LOH Group. Many local manufacturers build panels to the Rittal standard specifications. Protection is IP55 to EN 60 52 standards and complies with NEMA 12.

Modern cable trunking is open to prevent infestation and allow inspection and repair/modification (Figures 7.30 and 7.31).

Figure 7.30 Electrical panel enclosure with air conditioning unit.

Figure 7.31 Main trunking from panel and along the oven.

BIBLIOGRAPHY

Eratec, 2015. <www.era-tec.fr>.
Flynn Burner Corp, 2015. <www.flynnburner.com>.
Halifax Fan, 2015. <www.halifax-fan.com>.
Moro Srl, 2015. <www.moro.it>.
Omron Industrial Automation, 2015. <www.ia.omron.com>.
Rittal, 2015. <www.rittal.com>.
Rockwool, 2015. <www.rockwool.co.uk>.

CHAPTER 8

Oven Construction: Indirect Radiant Ovens

Contents

8.1 INDIRECT RADIANT BAKING CHAMBER

The following are the main considerations in specifying and installing an indirect radiant baking chamber:

- Baking chamber construction and dimensions
- Expansion joints
- Heater module

- Radiant tubes and ducts
- Extraction system
- Explosion relief
- Insulation
- Inspection doors
- Cleanout doors.

8.1.1 Baking Chamber Construction and Dimensions

The baking chambers are constructed in modules, normally about 2.3 m long. Each zone has a single heater module with a burner and combustion chamber. Materials for the baking chamber are mild steel 3.0 mm thick (Figures 8.1 and 8.2).

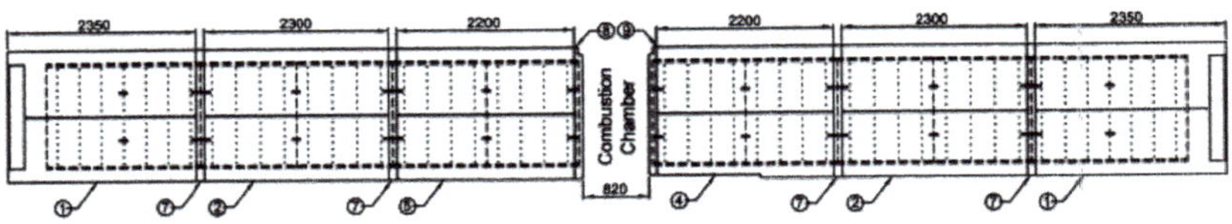

Figure 8.1 Typical zone layout for Baker Pacific radiant oven.

Figure 8.2 Baker Pacific indirect radiant oven installation of heater modules and baking chambers.

The baking chambers should be a minimum height (not exceeding 710 mm) and the radiant tubes should be no more than 215 mm from the band (top) and 165 mm (bottom) to ensure maximum intensity of radiation to the product. In order to ensure the radiant tubes are close to the oven band, the turbulence ducts are situated above and below the baking chamber (Figure 8.3).

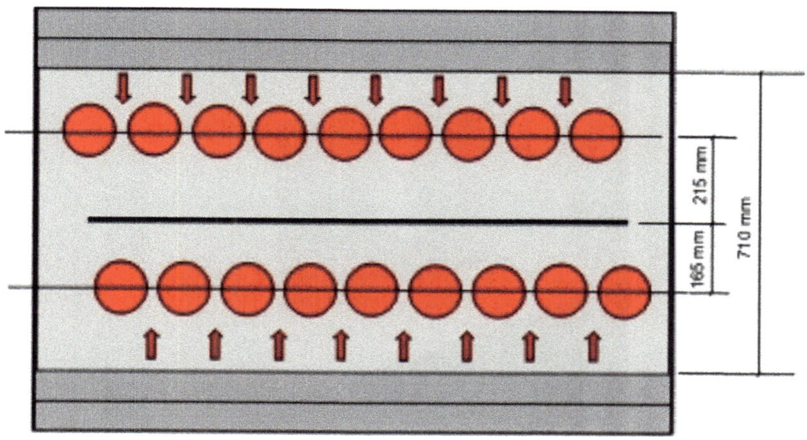

Figure 8.3 Section diagram for indirect radiant oven showing the position of the radiant tubes and the turbulence system ducts located at the top and bottom of the oven.

8.1.2 Expansion Joints

Expansion joints are located between each zone to allow expansion of the baking chamber.

These take the form of either stainless steel bellows, which open and close as the temperature of the baking chamber changes or a sliding male and female joint (Figures 8.4 and 8.5).

8.1.3 Heater Module

Each zone has a single burner. The burner fires into a stainless steel burner tube from where the hot gases enter a vertical duct at the back of the oven and are distributed to cross ducts at the top

Figure 8.4 Bellows type expansion joint.

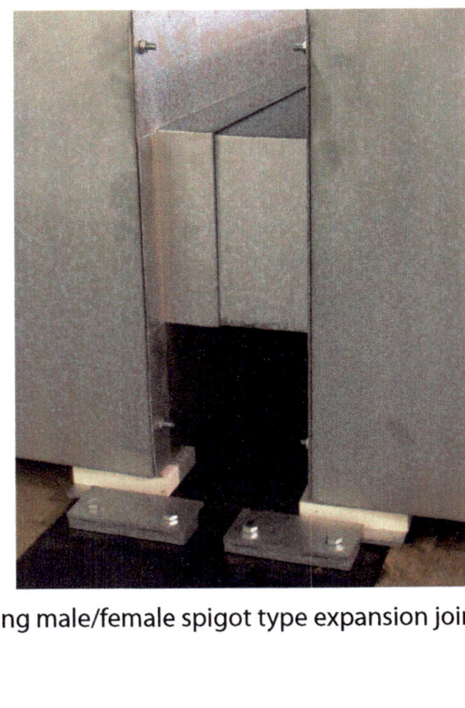

Figure 8.5 Sliding male/female spigot type expansion joint.

and bottom of the oven. From the cross ducts the gases enter the radiant tubes (Figures 8.6 and 8.7).

Figure 8.6 Heater module showing burner tube and distribution cross ducts, with adjustable plates for balancing lateral heat transfer.

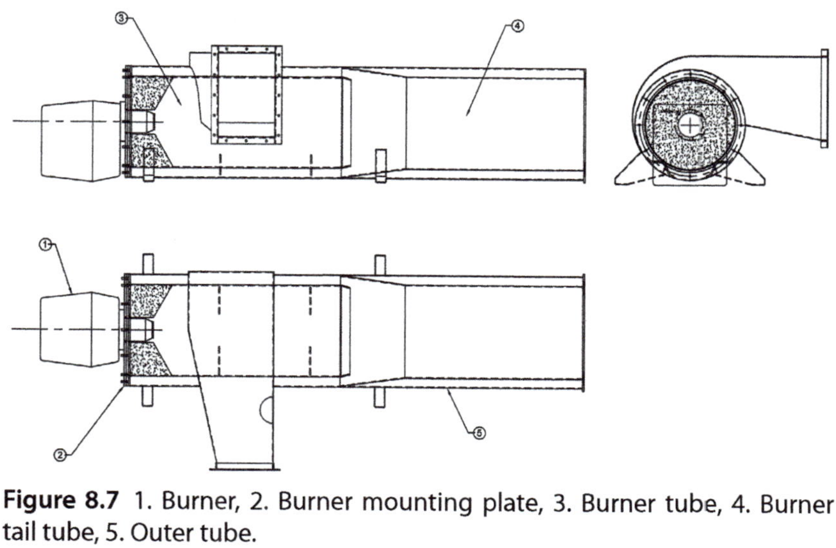

Figure 8.7 1. Burner, 2. Burner mounting plate, 3. Burner tube, 4. Burner tail tube, 5. Outer tube.

At the entry of the gases to the radiant tubes in the distribution ducts, sliding plates with adjustable blanking discs for each tube are located. These allow the heat distribution across the width of the oven to be balanced during commissioning. This facility enables the oven to achieve excellent lateral control of colour.

8.1.4 Radiant Tubes

Radiant tubes are manufactured from 4″ (101.6 mm) diameter mild steel tube. They are arranged along each zone with an offset, so that they do not run parallel to the band. This avoids the radiation from each tube to the products being applied to the products in parallel stripes (Figures 8.8 and 8.9).

Figure 8.8 Radiant tubes in an indirect radiant oven.

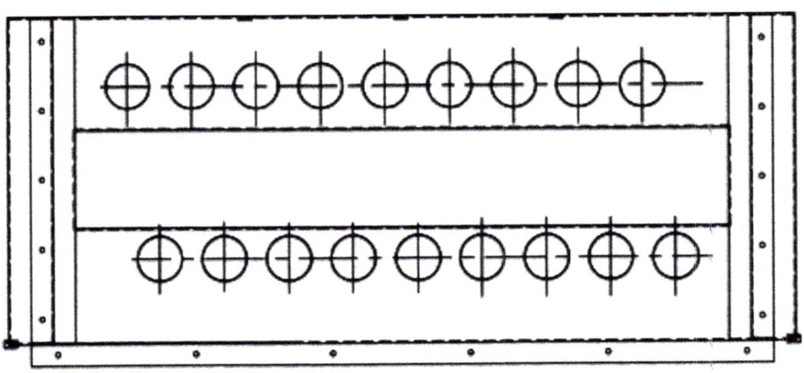

Figure 8.9 Note the offset of the radiant tubes.

The distance between the tubes and the oven band and the pitch of the tubes is important in maximizing the radiation intensity. Baker Pacific radiant tubes have a pitch of 145 mm. The top tube centres are 215 mm from the band and the bottom tube centres are 165 mm from the band.

8.1.5 Return Ducts

At the end of the zone the hot gases enter header ducts from where they are drawn through a return duct above the baking chamber, returning to the circulation fan.

8.1.6 Arrangement of an Indirect Radiant Zone

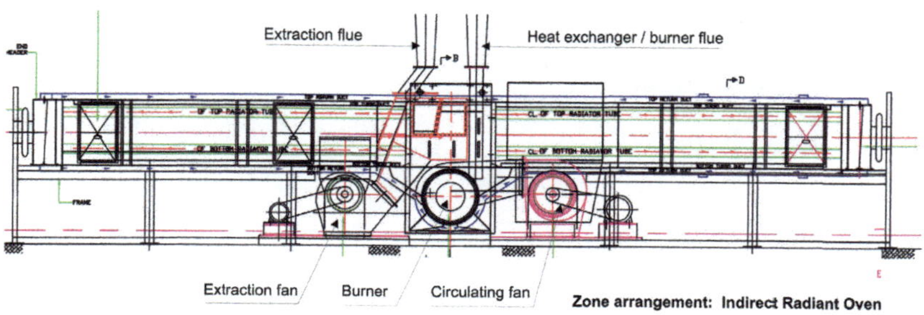

Figure 8.10 Drawing courtesy of Esspee Engineers.

8.1.7 Circulation Fan

The fan is located next to the burner. The fan draws the hot gases through the system: distribution ducts, radiant tubes, headers and return ducts and circulates the hot air back to the burner tube. Note that the radiant heating system is under negative pressure, which ensures that the products of combustion do not leak into the baking chamber (Figure 8.11).

8.1.8 Extraction and Turbulence

In each zone a duct above the band extracts the moist air to the extraction fan. From the fan the wet air enters a duct at the side of the oven where it may be directed to the turbulence duct or to the flue to atmosphere. The system is controlled by dampers (Figures 8.12 and 8.13).

Figure 8.11 Halifax 18P circulation fan for 1.25 m wide oven.

Figure 8.12 Halifax 15P extraction fan for 1.25 m wide oven.

Figure 8.13 The hot air from the baking chamber is extracted by the fan. It enters a duct above the fan and from there it can enter the turbulence ducts for top and bottom or it can be exhausted to atmosphere through the flue on the left. The damper positions are shown: the damper on the left controls the extraction to atmosphere and the two dampers on the right control the top and bottom turbulence system.

8.1.9 Heat Exchanger Flue (Chimney)

During operation the burner is drawing in air for combustion. In order to relieve the pressure, a flue with natural convection is arranged between the circulating fan and the burner tube.

This allows hot air and burnt gas to escape and maintains a negative pressure in the circulating system. The heat exchanger flue is on the right in Figure 8.10. This hot air and gas is relatively dry and is usually over 200°C. This is used in the heat recovery system.

8.1.10 Explosion Relief

In an indirect system, the risk of explosion is confined to the closed heating system. An explosion relief panel is located in the vertical duct opposite to the burner (Figure 8.14).

Figure 8.14 Back of heater module with cover removed to show position for explosion relief panel.

8.1.11 Insulation

The insulation is similar to that used for the direct gas-fired (DGF) ovens. Mineral wool slabs are used of 50, 100 and 150 mm thick and with 70 kg/m^3 density (standard) and up to over 100 kg/m^3 density for particularly hot areas. High temperature areas around the burner tube may be insulated by high temperature thermal ceramics, for example, Morgan Superwool Plus, density 128 kg/m^3, which has excellent performance up to 1000°C.

8.1.12 Inspection Doors

An inspection door with lamp is located in each zone (Figure 8.15).

Figure 8.15

8.1.13 Cleanout Doors

Cleanout doors are located at the back of the oven to give access to the baking chamber.

These are 'plug' doors which can be pulled out to give access for cleaning the baking chamber (Figure 8.16).

Figure 8.16 Large insulated cleanout doors are located in each baking chamber module behind the outer covers.

8.2 INDIRECT FIRED OVENS: BURNERS

8.2.1 Weishaupt Burners

Figure 8.17 Baker Pacific indirect radiant oven with Weishaupt WG burner installation.

8.2.2 Technical Description

The Weishaupt WG burner is a forced draught gas burner. For biscuit oven applications, the burner should always be a fully modulating type. The burner comprises the following features

- Combustion manager
    - Microprocessor control and monitoring of all burner functions
    - LCD display
    - Keypad operation
    - Data bus connection
    - Integrated valve proving of the solenoid valves.
- Flame sensor
    - Monitors the flame during operation. If a problem occurs, a safety lockout will operate.
- Double solenoid valve
    - Gas pressure governor. Controlled pressure is set by an adjusting screw

- Air/gas ratio control provides optimization over control range
- Two solenoid (Class A) valves
- Gas filter.

- Low gas pressure switch
- Gas pressure switch for automatic valve proving
- Air pressure switch

Loss of combustion air pressure activates a safety shut down (Figure 8.18)

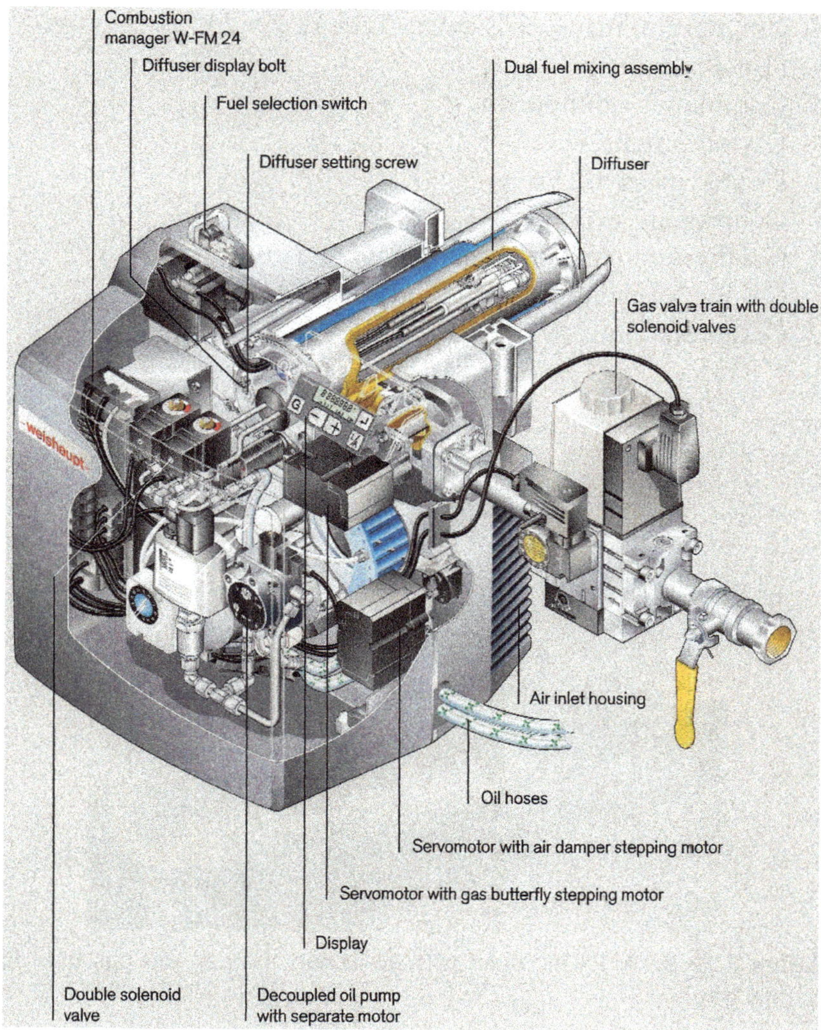

Figure 8.18 Picture courtesy of Max Weishaupt GmbH.

- Weishaupt gas train (Figures 8.19 and 8.20)
  Specification: Weishaupt burner WG30N/1-C ZM LN (for 1.25 m wide indirect radiant oven)
- Version: modulating via three-term switching and Buyer's PID controller
- Fuel: natural gas
- Gas pressure: 30 mbar
- Rating capacity: 40–350 kW
- Electro motor: 1 ph, 0.42 kW, 230V and 50 Hz
- Combustion manager: Siemens Type W-FM20
- Flame monitor
- Continuous running fan
- Gas valve train size 1″
- Double magnetic valve
- Gas pressure switch
- Gas pressure regulator assembly with safety valve
- Gas pressure inlet: (maximum) 2.5 bar
- Gas pressure outlet: 20 mbar

Figure 8.19 Baker Pacific oven with Weishaupt burner and gas train for natural gas.

Schematic of gas valve trains

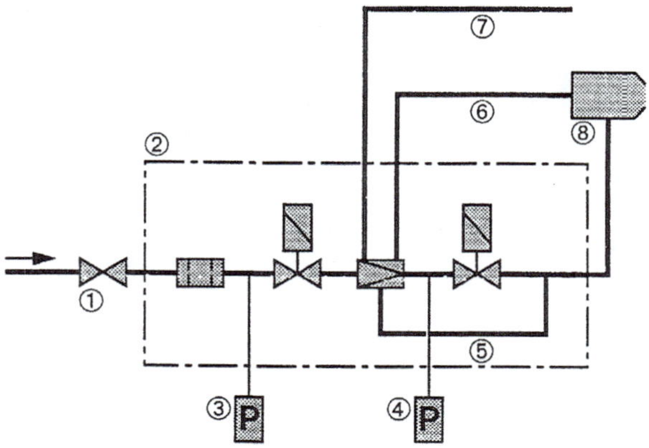

① Ball valve with thermal shutoff device
② Double solenoid valve DMV-VEF
③ Low gas pressure switch
④ Valve proving gas pressure switch
⑤ Gas pressure internal impulse line
⑥ Fan pressure impulse line
⑦ Combustion chamber pressure impulse line
⑧ Burner

Figure 8.20

- Gas filter
- Gas ball valve
- Fitting, reducing and sealing parts.

8.2.3 Maxon Burners

Figures 8.21 and 8.22

Specification for 1.2m wide indirect radiant oven Maxon OVENPAK 515 gas/oil burner:

Maximum heat release	1,250,000 Btu/hour (366 kW)
Minimum heat release	77,000 Btu/hour (22.5 kW)
Oil pressure required	60 psig (4.13 bar)
Natural gas pressure	2.0 bar (to be reduced to 0.005 bar)
Compressed air required	60 psig (4.13 bar)

Figure 8.21 Maxon OVENPAK 515SP dual fuel burner assembly.

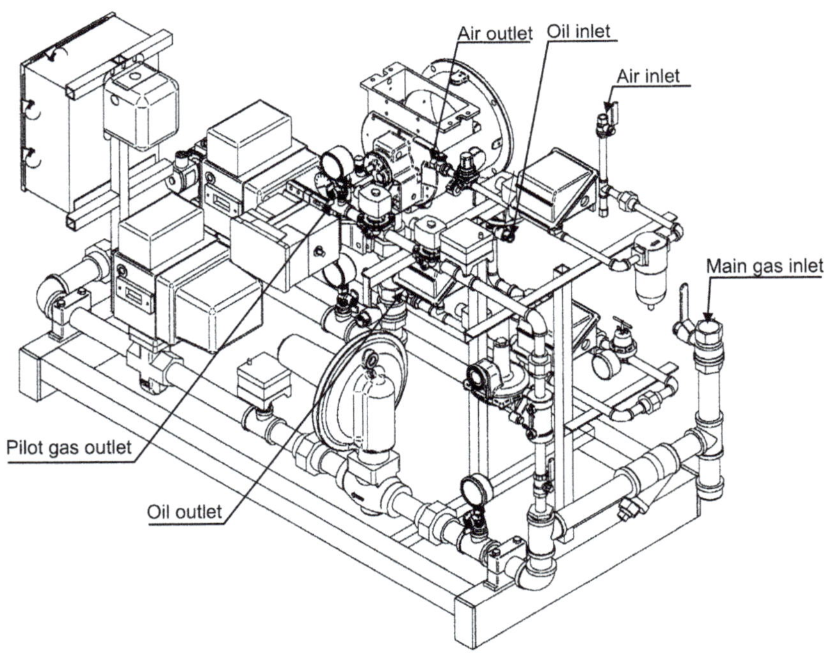

Figure 8.22 Maxon OVENPAK 515SP burner complete with gas feed train and oil feed train and compressed air train.

Each burner is complete with:

- Pilot adjustable orifice
- UV sensor device
- Dual type nozzle for oil/gas operation
- Integral combustion air blower
- Servo motor
- Ignition transformer
- Air pressure switch.

8.2.4 Maxon Gas Pipe Trains

- Gas strainer
- Gas cock
- Main gas regulator, size 1.25″
- Pressure gauge for upstream (0–30 psig)
- Pressure gauge for downstream (0–15″ WG)
- Low-pressure switch (0.5–4″ WG)
- High-pressure switch (5–28″ WG)
- Maxon solenoid valve series 5000
- Solenoid valve.

8.2.5 Pilot Gas Train

- Pilot regulator: high-pressure regulator
- Pilot solenoid: general purpose type
- Pressure switch for pilot
- Outlet pressure gauge
- Locking ball valve.

8.2.6 Oil Pipe Train

- Oil filter, size 3/8″
- Oil pressure regulator, size 3/8″
- Low-pressure switch
- High-pressure switch
- Solenoid valve for burner
- Locking ball valve
- Fuel: diesel oil.

8.2.7 Compressed Air Train

- Air filter
- Air pressure regulator
- Air pressure switch for low air pressure
- Solenoid valve for burner.

BIBLIOGRAPHY

Baker Perkins Ltd. <www.bakerperkins.com>.
Halifax Fan Ltd. <www.halifax-fan.com>.
Haas – Meincke. <www.haas.com>.
Weishaupt. <www.weishaupt-corp.com>.
Maxon Corporation. <www.maxoncorp.com>.
Morgan Advanced Materials. <www.morganthermalceramics.com>.

CHAPTER 9

Heat Recovery System

Contents

The heat recovery system (HRS) uses the waste heat from the burner flues. This may be used to heat one or two final zones of the oven. These zones would not require burners, giving a saving in capital and running costs.

All gas burners draw in a large amount of air for combustion. For complete combustion, 1.0 m^3 of gas requires 3.0 m^3 of oxygen (approximately 15 m^3 of air). This air is exhausted through the extraction system of a direct gas-fired oven and through the natural draught burner flue of an indirect oven.

The hot air and burnt gas in the burner flues of an indirect radiant oven are at a high temperature, typically up to 200°C in the first zones, and this hot air can be recovered and used for baking in a HRS.

A proportion of the hot gases in the burner flues are diverted to an HRS collection pipe, which runs along the top of the oven. Hot gases are collected from each zone with a burner. The hot gases are drawn along the collection pipe by a fan and blown into radiant ducts in the final oven zone (Figures 9.1–9.4).

Figure 9.1 Baker Pacific indirect radiant oven with heat recovery system.

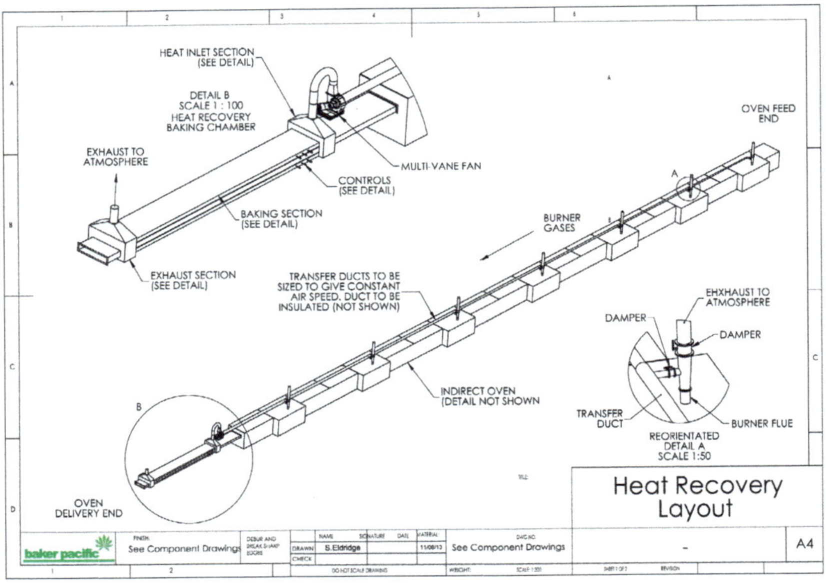

Figure 9.2

Figure 9.3

Figure 9.4

The burner flue (on the right) is connected to the HRS collection pipe, and the flow of hot gases is controlled by dampers. One damper controls the exhaust flue and one damper controls the connecting pipe from the burner flue to the collection pipe. These dampers are set by the commissioning engineer to allow sufficient quantity of heat for the final zone of the oven.

The HRS zone is constructed with radiant ducts above and below the oven band. The hot gases recovered from the burner flues are fed by the fan to the ducts. The fan is located on top of the oven at the end of the collection pipe (Figures 9.5–9.7).

Figure 9.5 Heat recovery zone.

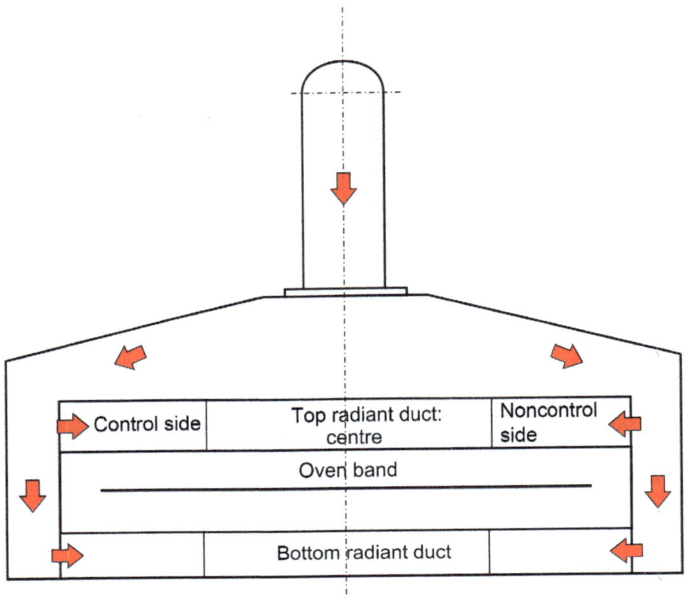

Figure 9.6

Figure 9.7 Control dampers for lateral heat balance.

The final zone of the oven has radiant ducts above and below the oven band. These ducts are divided along their length into three sections (control side, centre and noncontrol side). The flow of hot air into each section is controlled by a damper. These dampers may be adjusted to ensure the optimum heat balance top to bottom and across the width of the oven.

9.1 CALCULATIONS OF HOT AIR FLOW TO THE HRS ZONE

Calculations are based on the following oven specification:

1.2 m × 100 m indirect radiant oven

8 zones/7 burners × 350 kW/1 HRS zone

Product: Rotary moulded

Output: 3200 kg/hour

Energy used: 0.4043 kWh/kg

Total energy used: 0.4043 kWh × 3200 = 1294 kWh

Natural gas energy: 9.8 kWh/m^3

Gas consumption per hour = 1294/9.8 = 132 m^3

Air volume required per hour for complete combustion (approximately) 2000 m^3/hour

Total volume of products of combustion exhausted = 2000 m^3/hour

This is the total volume of hot burnt gas/air available for HRS.

Assume that we can use up to 50% of this volume: 1000 m^3/hour (0.28 m^3/s).

If we assume that the temperature of the hot air delivered to the final zone is 180°C, the energy available in the HRS zone can be calculated as follows:

Energy (kW) = air volume (m^3/s) × density of air (1.2 kg/m^3) × specific heat (1.005 kJ/kg °C) × temperature (°C)

Energy = 0.28 × 1.2 × 1.005 × 180 = 60.8 kW

The energy transferred to the HRS zone will be approximately 60 kW (maximum).

BIBLIOGRAPHY

The Engineering ToolBox. Air heating systems. Esspee Engineers, Kolkata, India. <www.engineeringtoolbox.com>.

CHAPTER 10

Oven Conveyor Bands

Contents

There are three major types of oven bands for conveying the dough pieces through the oven.

Each type of band has particular baking characteristics in terms of heat transfer and physical performance, and therefore has a particular range of applications.

10.1 OPEN WIRE-MESH BANDS

These are the most versatile and widely used oven bands. They are suitable for crackers, hard biscuits and rotary moulded products, in fact all types of biscuits except soft cookies with high fat contents, where the dough will flow into the holes in the mesh. The wire-mesh bands allow convective air to pass through, heating the base of the dough pieces (Figure 10.1).

Figure 10.1 Rolled wire-mesh band from Steinhaus GmbH.

These wire-mesh bands are normally made from carbon steel wire coils which are linked together. The links give a flat surface to the band and good flexibility. Alternative materials are available for special applications, for example, stainless steel wire-mesh bands may be used where the baking process has very high humidity for products such as cheesecake.

The most popular bands are often referred to as Z47 and Z47R. These designations are from a European supplier, Produits Trefiles de la Bridoire (Agrati Group). An equivalent high-quality wire-mesh band is type F4102 from Steinhaus Gmbh. The high-quality balanced weave belts from the United States are from Cambridge Engineered Solutions and Ashworth Bros Inc. (Figures 10.2–10.7).

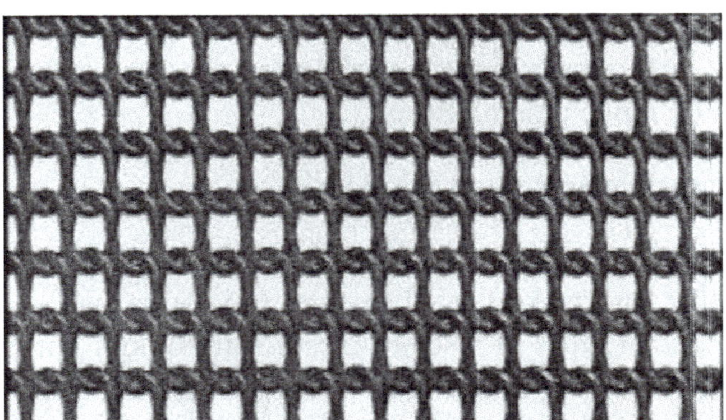

Figure 10.2 Z47 from La Bridoire Wire thickness 1.2 mm.

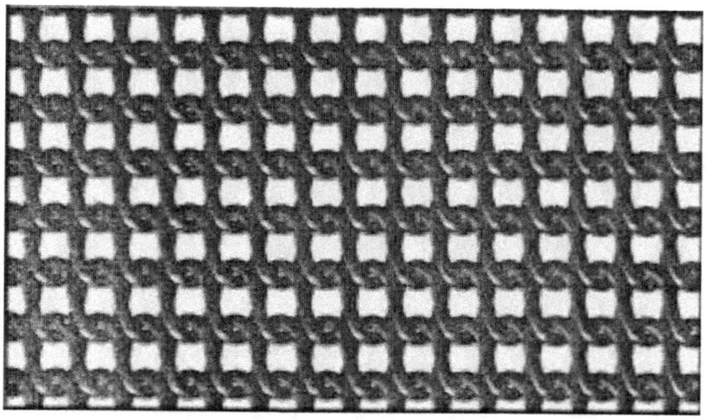

Figure 10.3 Z47R from La Bridoire Wire thickness 1.5 mm.

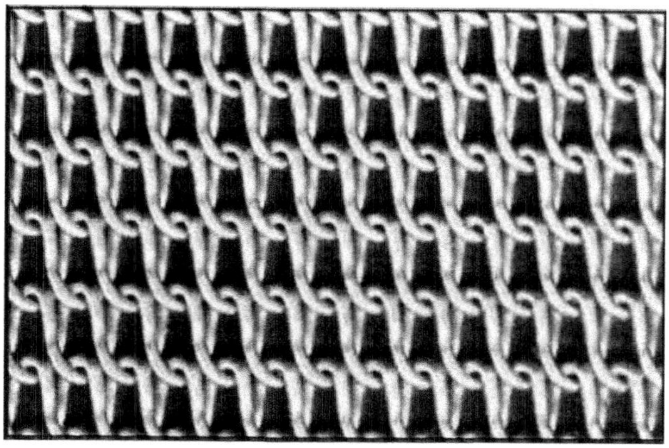

Figure 10.4 Steinhaus F4012 belt (equivalent to Z47 type).

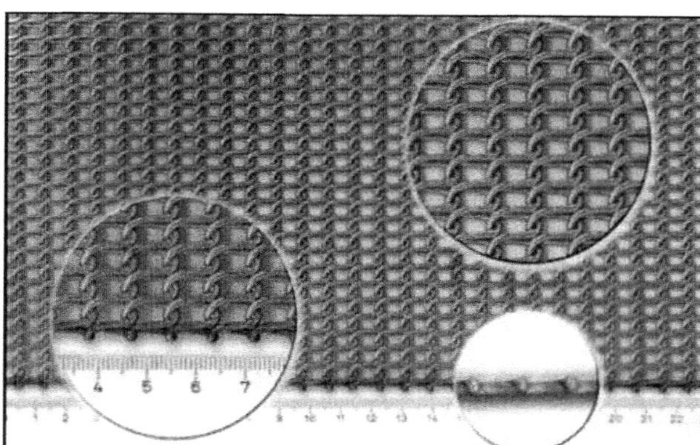

Figure 10.5 Steinhaus F4102. Even mesh structure and angularity, accurate width welded edges.

Figure 10.6 Cambridge Engineered Solutions: balanced weave belts.

Figure 10.7 Ashworth U66-48-1718 PGLW.

10.1.1 Wire-Mesh Bands: Skid Bar Supports

Wire-mesh bands are successfully supported in the baking chamber by cast iron skid bars. These bars are placed inside the baking chamber with a pitch of 0.8–1.2 m. They are supported in frames and are removable for cleaning. Apart from cleaning, which is infrequent (normally annually), skid bars do not require any maintenance (Figure 10.8).

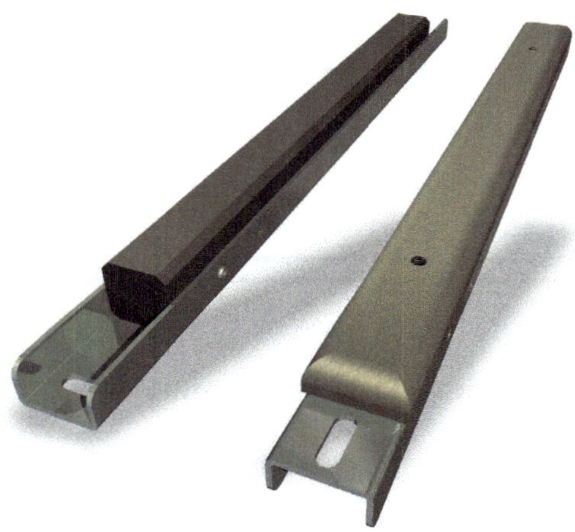

Figure 10.8 Sandvik Process Systems graphite and cast iron skid bars.

The skid bars are cast from iron with a graphite content which reduces friction. The material and casting are critical to the performance and the skid bars are supplied to suitable specifications, for example, International ISO/DIS 185 Grade 200/250. The chemical composition, casting process, shape of the mould and cooling rate are important to achieve the optimum low friction and durability of the skid bars in operation.

The surface of the skid bar has a microstructure of a pearlitic matrix with evenly spread type A graphite flakes to ASTM A247 standard with no hard particles. Maximum allowed amount of ferrite is 10%. No ledeburite is allowed. The sliding surface of the skid bar is machined after casting by removing the surface layer (1.0 mm), which contains ferrite and would result in scratching of the band. The final surface should have a roughness of Rmax 3.0 μm.

The installation of the skid bars is critical to good band tracking. The bars must be level and at right angles to the direction of band travel. Levelling is best done optically with a laser instrument (Figure 10.9).

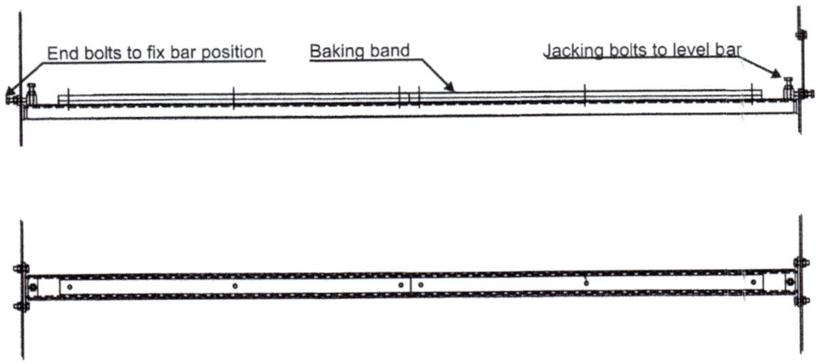

Figure 10.9 Baker Pacific skid bar support with levelling adjustment.

10.1.2 Return Band Supports

On the return below the baking chamber, the band is supported on rollers. A proportion of these, usually one in three, may have spring-loaded guide discs. These are very effective in keeping the band in its central position (Figure 10.10).

Figure 10.10 Baker Pacific return band support roller with spring-loaded guides.

10.1.3 Wire-Mesh Oven Band Cleaning

Oven band cleaning is critical to the quality of the biscuit and the oven performance. We strongly recommend a band cleaner which operates all the time during baking, ensuring that there is no build-up of carbon, which is difficult to remove.

The Baker Pacific oven band cleaner consists of two independently driven wire brushes for cleaning both sides of the band. The unit is designed for a continuous, gentle cleaning action to maintain the band in good condition. The top wire brush, driven by a fixed speed motor gearbox, cleans the inside of the oven band. The lower wire brush unit cleans the outside of the oven band. It is mounted on a trolley for removal to the side of the oven. Guide rails are provided for the trolley with stops and locking handle. The brush is raised or lowered to the working position by a lever and the brush pressure is adjustable (Figure 10.11).

Figure 10.11 Baker Pacific oven band cleaner.

Maintaining a clean band is important, particularly when baking products with sugar topping.

Preventing contamination is vital. When baking products with toppings a good system of recovery for surplus toppings is essential to prevent toppings dropping on the oven band. A long wire-mesh panning conveyor is recommended with a level transfer of the dough pieces to the oven band. The panning web should be an open mesh band (enrober wire-mesh) and be around 2–3 m long. The end transfer roller or nosepiece is of small diameter to allow a level transfer of dough pieces to the oven band.

For wire-mesh bands which have become dirty, some hard carbon deposits may be removed by heating the band to 400°C, when the hard carbon deposits will break and can be removed by the brushes. It is important that the band is heated evenly, and only sufficient temperature to carbonise the debris is used. Uneven or excess heating will distort the band.

Soft deposits may be removed by steam cleaning with an industrial cleaning fluid. Suitable drainage is required. After cleaning, the band should be dried and oil (soybean oil/coconut oil) applied to prevent rusting.

10.1.4 Wire-Mesh Oven Band Tracking

The position of the oven band edge should be constantly monitored by vertical rollers on each side. The rollers are supported on a bar which follows the movement of the band. The bar carries a Euchner trip rail which engages a multiple limit switch. If the band moves up to 10 mm, it will automatically actuate a pneumatic cylinder which moves a tracking roller below the band. The oven band will always try to leave a support roller at 90 degrees to its angle.

Therefore by altering the angle of the tracking roller, the band is pushed back to its central position. If the band moves more than 25 mm, then the drive will automatically stop. These settings are adjustable (Figures 10.12 and 10.13).

The tracking assemblies are mounted just before each end drum to guide the band onto the drum in a central position. The delivery end tracker is on the top and the feed end tracker is on the return band at the bottom of the oven. The trackers may have several rolls on a single frame operating together (Figure 10.14).

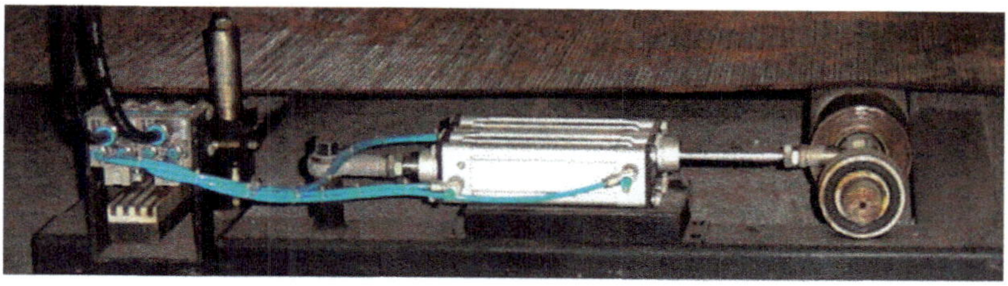

Figure 10.12 Euchner trip rail and limit switch with pneumatically operated tracking roll.

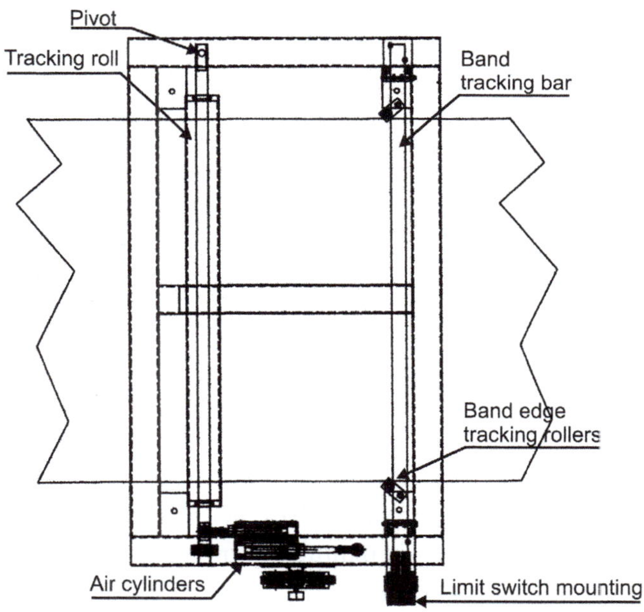

Figure 10.13 Baker Pacific auto tracker assembly.

Figure 10.14 Band tracker at delivery end of oven.

Vertical guide rollers are also used inside the baking chamber. Typically there is one assembly in each oven zone adjacent to the inspection door. These rollers are pivoted and may be spring loaded (Figure 10.15).

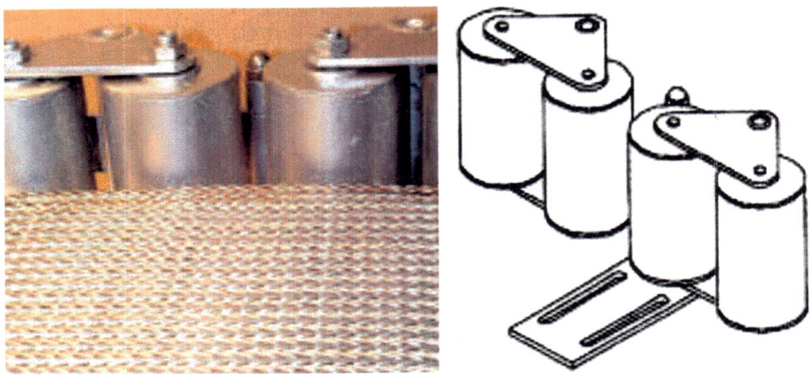

Figure 10.15 Guide rollers by Ashworth Bros, suitable for all spiral meshes.

An alternative method of band tracking moves the delivery end drum to skew it. It is pivoted at one side and moved by a pneumatic cylinder at the other side to alter its angle to the direction of travel of the band. However, this system is not recommended by some band manufacturers as it alters the distance between the feed end drum and the delivery end drum on one side, creating a tight side and a loose side of the band. Oven band manufacturers generally recommend that the drums are set exactly at right angles to the direction of travel of the band.

10.1.5 Joining Wire-Mesh Bands

To join the ends of a wire-mesh band, two spiral joining wires are used. These are inserted from the side of the band. After insertion, the spiral joining wires are hammered flat to provide a secure and almost invisible joint (Figure 10.16).

Figure 10.16 (a) Inserting a spiral joining wire, (b) flattening the joint, (c) the final joint is almost invisible. *Pictures from Steinhaus GmbH.*

10.2 HEAVY MESH BANDS

These are tightly woven bands with a 'herring bone' pattern presenting a continuous flat surface. They are relatively heavy and have a high heat mass. These bands are preheated and will conduct heat directly into the base of the dough pieces. They are widely used, particularly in North America, for the baking of soda crackers, but are also versatile and may be used for all types of biscuits, except high-fat cookies (Figures 10.17–10.19).

Figure 10.17 Cambridge Engineered Solutions: compound balanced weave belt.

Figure 10.18 Ashworth CB5 Baking Band. Ashworth CB5 27-84-1416F.

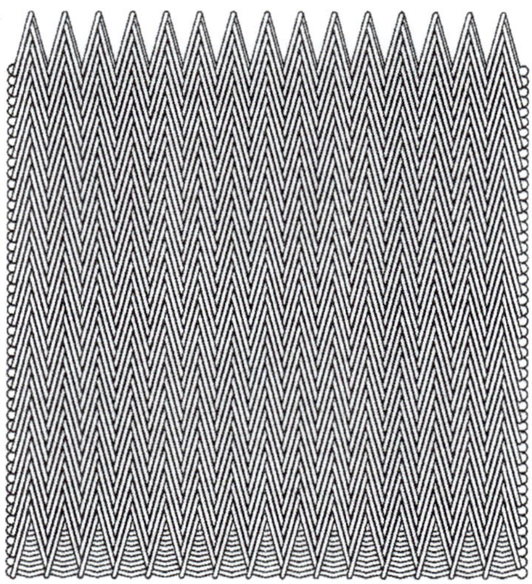

Figure 10.19 Ashworth CB5 Baking Band. Ashworth CB5-36-120-18.

10.2.1 Heavy Mesh Band Supports

Heavy mesh bands are supported on rollers, which will carry high band loadings with low friction.

The design of the supports and adjustment for levelling and the specification of the bearings are critical as the rollers operate at baking temperatures. If the rollers seize, they will quickly cause wear to the oven band. The ends of the rollers are therefore exposed for inspection and high-temperature bearings must be used. The bearings are mounted outside the baking chamber and are insulated from the chamber (Figure 10.20).

Figure 10.20 Bearing assembly with SKF bearing type Y – V228 suitable for operating temperatures up to 350°C.

10.2.2 Heavy Mesh Band Tracking

Ashworth bands are tracked by vertical side rollers positioned before each drum. These rollers are not intended to force the band into position. They do limit side movement and act as sensing devices to indicate tracking problems.

The band control units should be located at each end of the oven before the terminal drum.

The distance between the control unit and the drum should be three times the band width. At this distance the force required to move the band to its central position on the drum is small (Figures 10.21 and 10.22).

10.2.2.1 Ashworth Control Unit

- *Base frame with three-band support rollers*
- *Four vertical rollers on each side of the band*

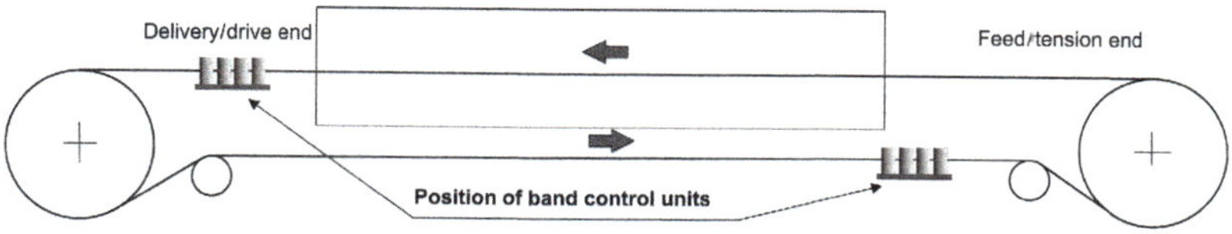

Figure 10.21 Recommendation for position of control units.

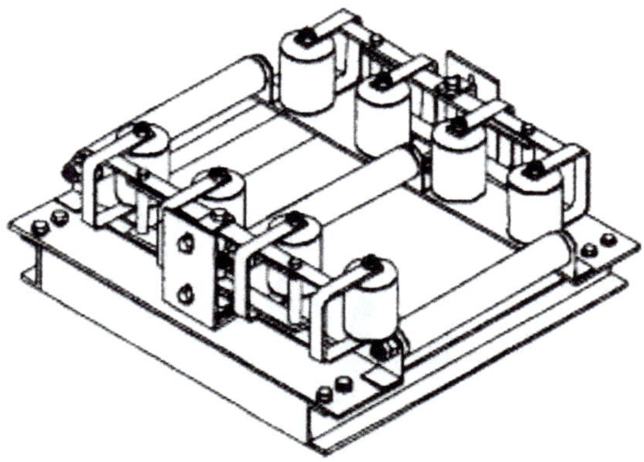

Figure 10.22 Ashworth control unit.

- *Band width up to 1525 mm*
- *Suitable for all spiral wire-mesh bands*
- *Guide rollers adjust vertically to move wear point*
- *Available with carbon steel ball bearings for temperatures up to 350°C or with carbide zero wear bearings for higher temperatures, no lubrication required.*

10.2.3 Joining Ashworth Bands

The bands are joined by inserting the correct number of connecting wires, depending on the type of mesh. The crimped connectors are inserted and seated so that the weave is flat. The connecting wires extend each side by 2 mm and are fastened to the end spirals by a simple weld (Figures 10.23–10.26).

Note: Extensive technical data is available at www.ashworth.com.

Figure 10.23 CB5 band: five connectors.

Figure 10.24 CB3 band: three connectors.

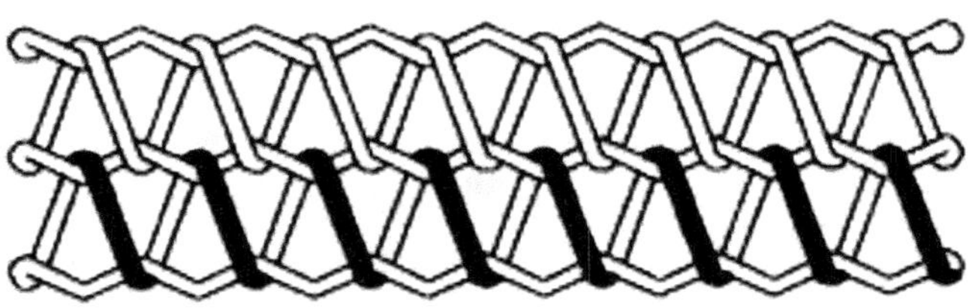

Figure 10.25 Unilateral weave: one connector.

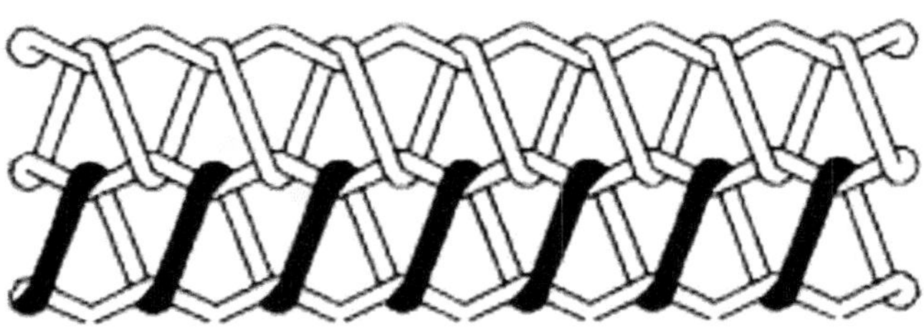

Figure 10.26 Balanced weave: one connector.

10.3 STEEL BANDS

These bands are made from thin (usually 1.2 mm) carbon steel and are suitable for soft doughs, which flow on the band when heated. They are used for all deposited cookies and cakes, such as layer cake, which are baked directly on the oven band. Traditionally steel bands are also used for baking 'Marie' biscuits (Figure 10.27).

Figure 10.27 Cookies deposited directly onto a steel band.

10.3.1 Steel Band Supports

Steel bands are successfully supported on skid bars as described above. Skid bars are manufactured and recommended by Sandvik Process Systems.

In the case of steel bands, the skid bars are 75–100 mm narrower than the band. The pitch of the skid bars should be 0.8–1.0 m.

Special soft graphite skid bars may be used at the feed end of the oven. Approximately four graphite skid bars are used for a

30m long baking chamber. These skid bars lubricate the inside of the band and reduce friction. The soft graphite skid bars apply graphite in an even, continuous, automatic way. They provide the following advantages:

- There are no limitations on speed and they are ideal for heavily loaded bands
- The band does not need additional lubrication at start up or during operation
- Band life is increased
- Can be used in conjunction with cast iron skid bars to help reduce friction and eliminate any slip-stick problems.

Hard graphite skid bars are also available. These provide a long life at high temperatures. They may be used in place of cast iron skid bars.

10.3.2 Joining Steel Bands

Steel bands may be riveted, and Sandvik Process Systems supply riveting kits including the tools and instructions. However, a much better method is welding, which provides a completely smooth surface to the join. Welding is a specialist operation carried out by the band supplier with specialist equipment for welding and tempering the joint. Sandvik Process Systems provide equipment and specialist engineers.

10.3.3 Steel Band Cleaners

Steel bands may be cleaned by rotating brushes (as described in Section 10.1.3 for the wire-mesh bands) assisted by scrapers.

For soft, sticky products, such as layer cake, steel scrapers are used. Up to three rows of spring steel scrapers in a staggered formation are mounted on the adjustable frame. The frame is raised by a lever to scrape the band. The pressure is adjustable (Figure 10.28).

For very sticky products hot water washer units may be used. These are located close to the delivery end of the oven, where the band is still hot. The washer may be used continually or intermittently as required. The water is sent to drain. After washing, the band should be dried and oiled to prevent rust.

Figure 10.28 Steel band scrapers on a cake oven.

10.3.4 Steel Band Greasing

At the feed end drum, the band may be greased to prevent the dough sticking. The grease is deposited from a hopper onto soft pads which are mounted on a reciprocating arm. This moves across the band, back and forwards to smear an even coating of grease on the steel band (Figure 10.29).

Figure 10.29 Feed end of a layer cake oven with band greaser and depositor.

10.3.5 Steel Band Tracking

Spring-loaded guide rollers provide a robust and safe way to keep a band running within acceptable side movements. A pair of side guide rollers may be installed in each oven zone (Figures 10.30 and 10.31).

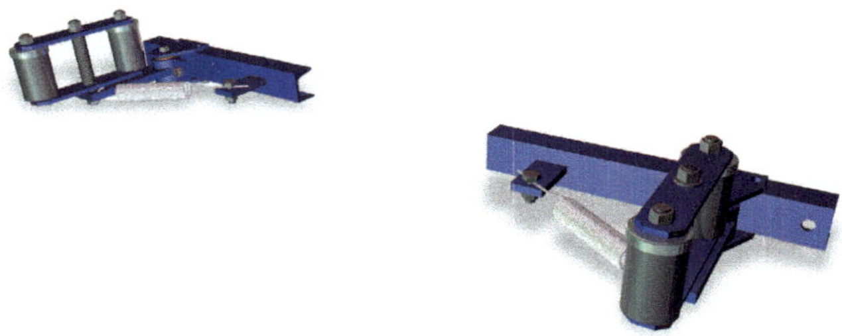

Figure 10.30 Spring-loaded side guide rollers from Sandvik Process Systems.

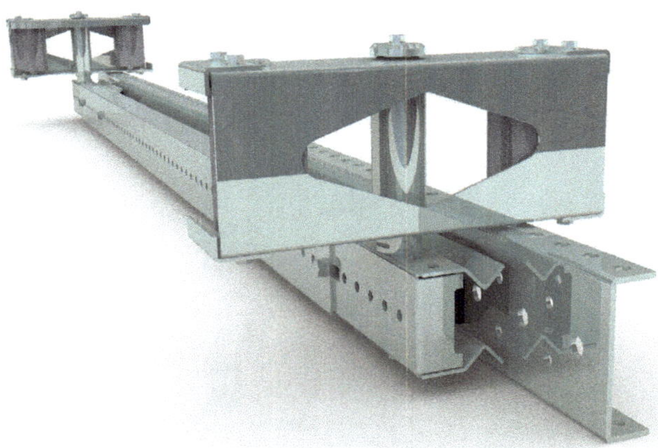

Figure 10.31 Spring-loaded guide rollers from Sandvik Process Systems.

Steel bands may be tracked with tilting rollers activated by a band edge detection device. The tilting rollers are automatically raised on one side, creating a tight (higher tension) and a slack (lower tension) side to the band. The band will track towards the side with the lower tension (Figure 10.32).

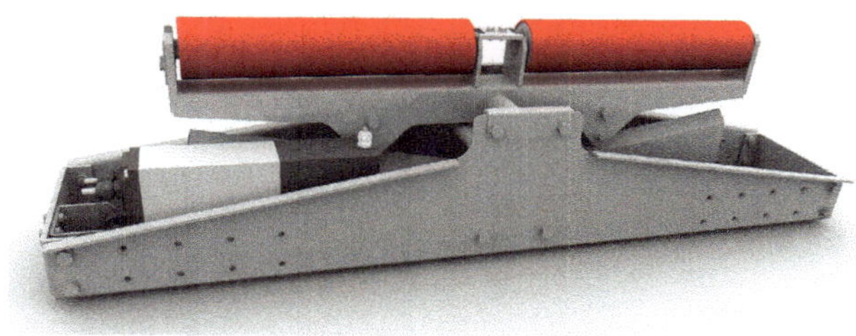

Figure 10.32 Compact belt tracker from Sandvik Process Systems.

BIBLIOGRAPHY

Ashworth Bros., Inc, 2014. Balanced weave belts, cleaning baking bands, control systems, baking bands – installing. <www.ashworth.com>.

Baker Perkins Ltd., Cleaning wire-mesh bands. <www.bakerperkinsgroup.com>.

Berndorf, 2014. Bake oven belts. <www.berndorf-band.at>.

Cambridge Engineered Solutions. www.cambridge-es.com.

Euchner GmbH & Co. KG. www.euchner.co.uk.

Jan-Ola Jonsson, The Properties of Steel Belts.

Produits Trefiles de la Bridoire, 2015. Wire bands. <www.agrati.com>.

Sandvik Process Systems, 2015. Product information. <www.smt.sandvik.com>.

SKF, 2015. Product information. <www.skf.com>.

Smith, W.H., 1967. Cleaning and Care of Oven Bands. Biscuit Maker and Plant Baker, London.

Steinhaus Gmbh, 2015. Rolled baking oven belts. <www.bakingovenbelts.com>; <www.steinhaus-gmbh.de>.

CHAPTER 11

Oven Conveyor Design

Contents

The products are carried through the oven by a conveyor which has a turning drum at each end of the oven (Figure 11.1).

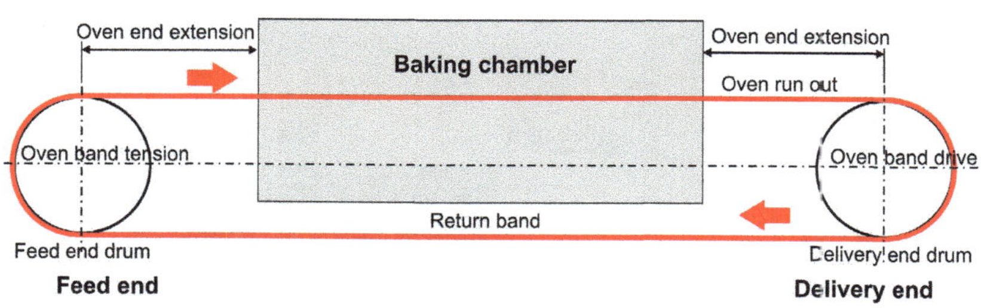

Figure 11.1 Diagram of oven band circuit.

11.1 FEED END

The feed end unit has the following functions:

- Oven band drum support and movement
- Oven band tension
- Tracking
- Installation of equipment over the oven band (eg, a wire-cut machine) (Figures 11.2–11.4).

Figure 11.2 Baker Pacific Feed End on a cracker oven.

Figure 11.3 Baker Pacific Feed End on a steel band oven with depositor.

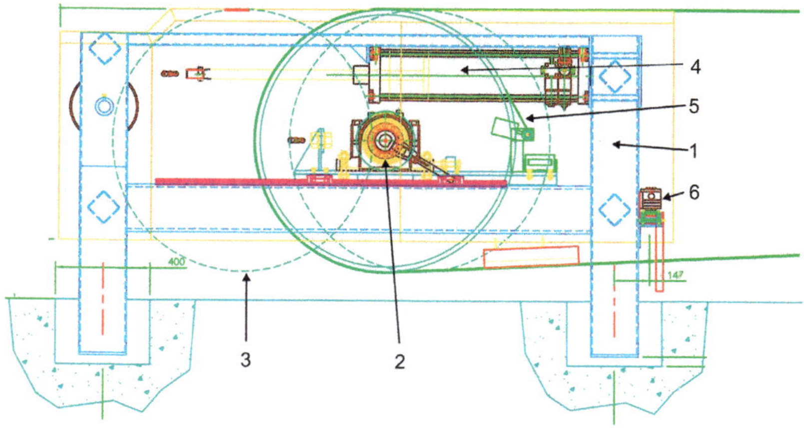

Figure 11.4 Feed end assembly drawing showing: (1) rigid structure anchored in the floor, (2) oven drum supported on bearings mounted on slides, (3) oven drum movement for tensioning the band, (4) pneumatic cylinders for tensioning, (5) oven drum scraper, (6) band position detector.

11.1.1 Oven Terminal Drum

The oven terminal drums support the band as it turns. The drums are manufactured to high quality and close tolerances to ensure good band tracking. The tapered edges help to hold the band in position on the drum (Figure 11.5).

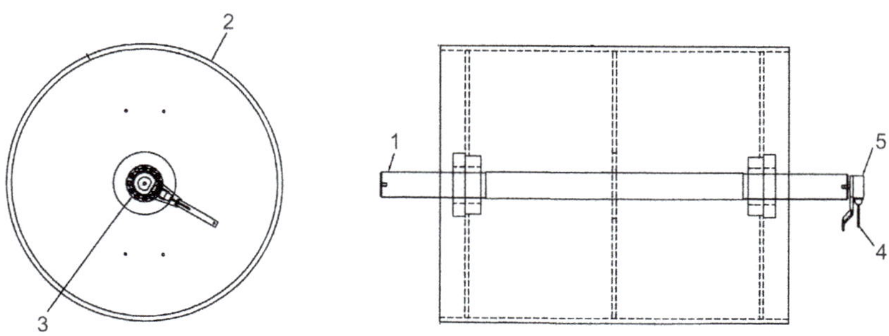

Figure 11.5 Oven terminal drum for 1250-mm wide wire-mesh band (total width 1350 mm): (1) shaft, (2) machined drum 1100 mm diameter with taper of 2 mm at each side, (3) Cross and Morse clamping element, (4) proximity switch, (5) Synatel speed monitor.

The drum is secured to the shaft by special clamping elements. These elements use frictional forces to transmit high torque. Precision-tapered thrust cones create high pressure between the shaft and the drum hub to secure the drum. Large clamping surfaces ensure the pressure level is not excessive. The clamping elements are auto centering (Figure 11.6).

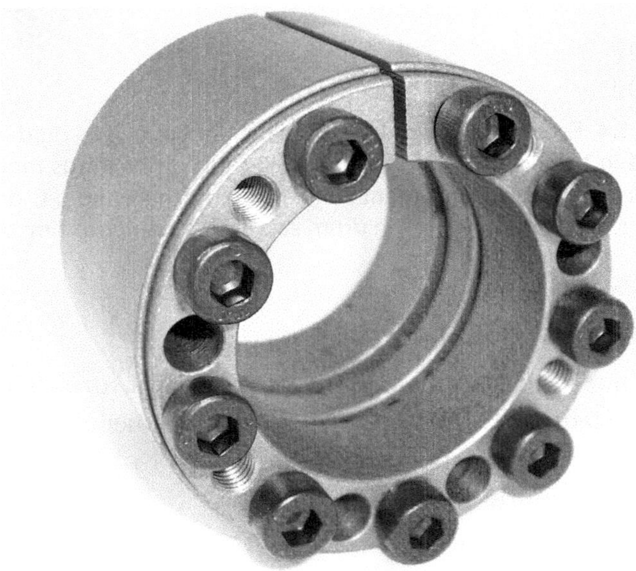

Figure 11.6 Cross and Morse clamping element RCK11.

The drum shaft pillow block bearings are mounted on precision slides to allow the drum to move to provide the band tension. The drum must move exactly at right angles to the direction of band travel and so the rigidity of the feed end structure and the slides are critical (Figure 11.7).

The tension is applied to the bearing blocks by chains pulled by pneumatic cylinders mounted on each side of the feed end unit (Figures 11.8–11.10).

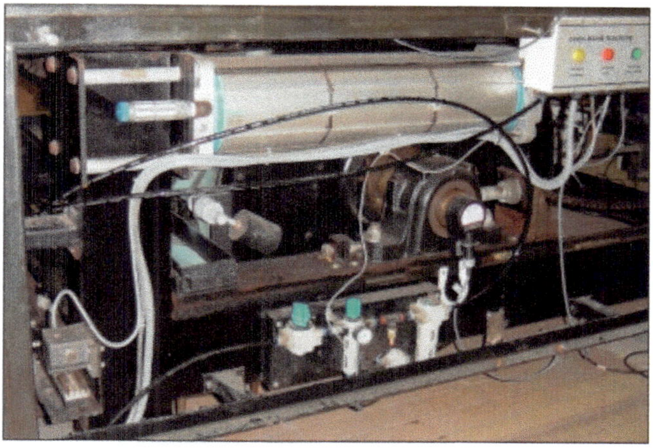

Figure 11.7 Feed end assembly with pneumatic cylinder, drum bearing with Synatel speed monitor, drum scraper, band position detector.

Figure 11.8 Star linear ball rail slides from Rexroth Bosch.

Figure 11.9 Martonair pneumatic cylinder for tensioning the band.

Figure 11.10 Synatel speed monitor for drum rotation.

11.2 DELIVERY END

The delivery end unit has the following functions:

- Oven band drum support
- Oven band drive
- Tracking
- Stripping knife and conveyor
- Installation of band cleaner
- Oven end hood and extraction fan (Figures 11.11–11.13).

Figure 11.11 Baker Pacific oven delivery end with oven end hood and fan, band cleaner, band tracker with band position monitor and pneumatically operated tracking roll.

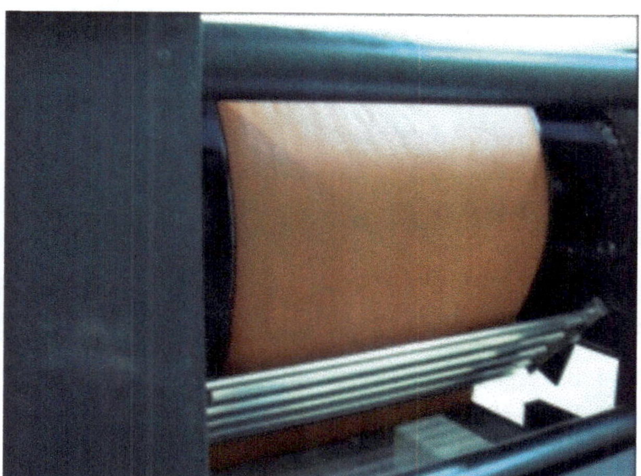

Figure 11.12 Delivery end drum on a Baker Pacific oven producing layer cake.

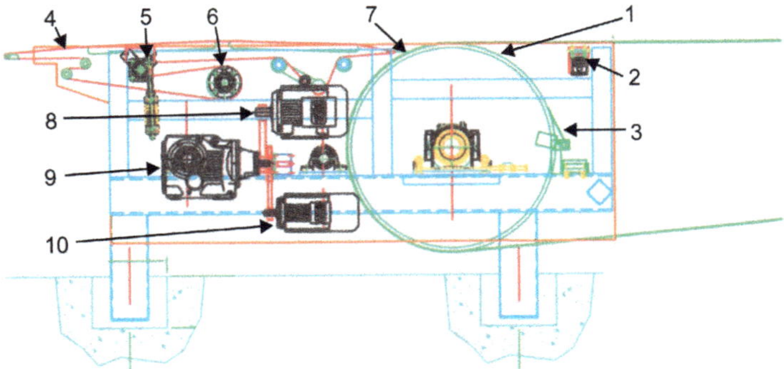

Figure 11.13 Delivery end arrangement drawing: (1) oven end drum, (2) band position detector, (3) oven drum scraper, (4) stripping conveyor, (5) biscuit reject actuator, (6) stripping conveyor drive roll, (7) stripping knife, (8) main motor 7.5 kW, (9) reduction gearbox, (10) emergency DC motor.

11.2.1 Delivery End Drum

The drive shown above is for a 1.25 m × 100 m biscuit oven. The main motor is 7.5 kW with a reduction gearbox. The DC motor is powered by batteries and operates automatically in the event of a power failure (Figures 11.14–11.16).

Figure 11.14 Baker Pacific delivery end drum, band position detector, drum scraper, main and DC motors. The drum design and construction is the same as the feed end drum.

Figure 11.15 Drum scraper.

Figure 11.16 Main drive and emergency DC motor with sprag clutch.

11.2.2 Oven Drive

The oven is driven from a 7.5-kW AC motor through a toothed belt drive to a reduction gearbox. The gearbox is mounted on a 60-mm diameter shaft, which transfers the drive to the opposite side of the delivery end unit. The speed is further reduced by the chain and sprocket transmission. The chain drive is also used to drive the stripping conveyor. In this mode the DC motor is idle due to the orientation of the sprag clutch. The motor is controlled by inverter to provide a speed range of 3.33–33.3 m/minute (corresponding to a 3-minute baking time).

If a power cut occurs, the DC motor is automatically engaged. The power is transmitted by a toothed belt drive to the reduction gearbox. In this mode the AC motor is idle. The DC motor drives the oven band at 6.44 m/minute to empty the oven of dough pieces (Figure 11.17).

Figure 11.17 Chain and sprocket drives to the drum and to the stripping conveyor.

11.2.3 Sprag Clutch

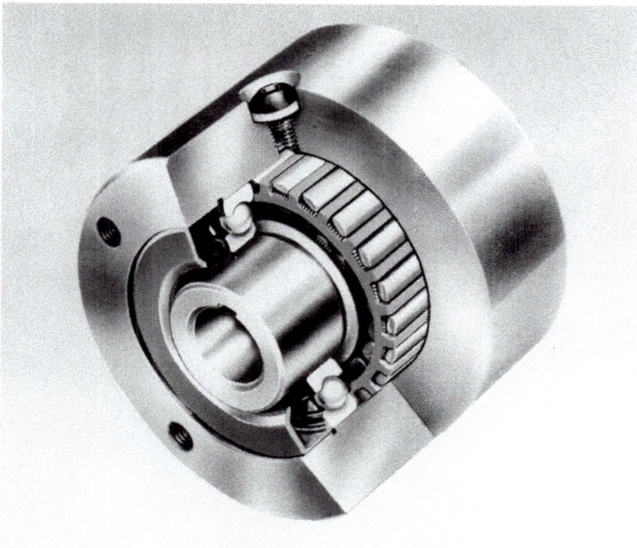

Figure 11.18 The sprag clutch is a one-way free wheel clutch suitable for over-running. It has an inner and an outer race with precision ground cams in contact with both races. The clutch drives both the main drive and the emergency drive. The unit shown is a Cross and Morse MO500 clutch.

11.2.4 Uninterruptible Power Supply

The oven requires an emergency drive system to operate automatically in the event of a power failure. This may be a factory generator which automatically takes over the power supply and supplies the oven drive. Alternatively a DC motor and battery power may be used (Figures 11.19–11.21).

The battery pack is continuously charged during normal operation. In the event of a power failure, the batteries will power the DC motor at 2.55 kW for 15 minutes to empty the oven of biscuits. The band speed will be 6.44 m/minute.

Figure 11.19 AC motor at top and DC motor below.

Figure 11.20 Erskine battery charger.

Figure 11.21 Erskine battery pack for 1.2 m × 100 m Baker Pacific oven.

11.2.5 Stripping Conveyor

The biscuits are stripped from the oven band by a knife and transferred to the stripping conveyor. This is usually a wire-mesh conveyor driven by the main oven drive (Figures 11.22–11.24).

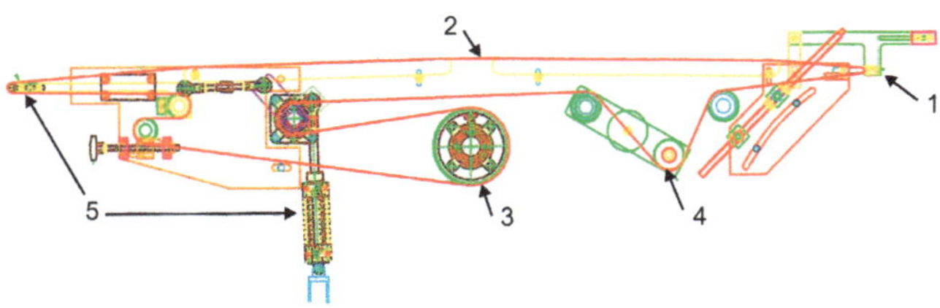

Figure 11.22 Stripping conveyor arrangement: side elevation. (1) Stripping knife at oven end, (2) stripping conveyor, (3) drive roll, (4) weighted tension roll, (5) biscuit reject mechanism.

Figure 11.23 Stripping knife: transfer from oven band to stripping conveyor.

Figure 11.24 Transfer from stripping conveyor to cooling conveyor.

Burnt or broken biscuits may be rejected at the end of the stripping conveyor by operation of an air cylinder which retracts the end nosepiece of the conveyor. The biscuits will fall onto a narrow cross conveyor which will automatically start and will deposit the scrap biscuits into a bin at the side of the line (Figure 11.25).

Figure 11.25 Cross conveyor to collect scrap biscuits from the reject.

11.3 OVEN END HOOD DESIGN

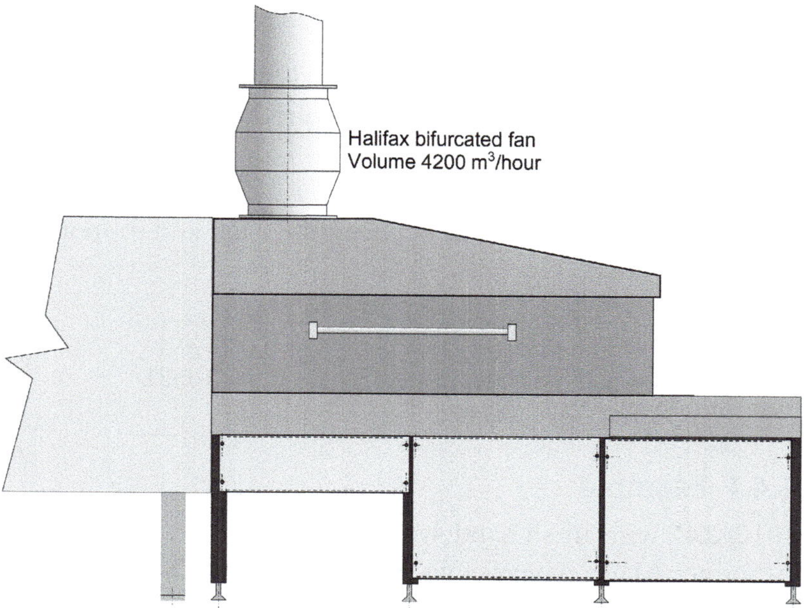

Figure 11.26 Recommended design: low hood, fully enclosed at sides and bifurcated fan close to the oven end.

Figure 11.27 Baker Pacific oven end hood.

11.4 CALCULATION OF OVEN BAND TENSION

The formula is $T = (WLf_r + wLf_r)$

where

T = Band tension in kg/m of band width

W = Total weight = Band weight + product weight in kg/m^2

L = Conveyor length in m

w = Band weight in kg/m^2

f_r = Coefficient of friction between the band and supports

Typical f_r values:

Cast iron skids (dry)	0.23–0.4
Cast iron skids (lubricated)	0.13–0.21
Rollers (free turning)	0.10

11.4.1 Example

F4012/Z47 wire-mesh band: weight 7.3 kg/m^2

Product: Marie type: weight of dough pieces 1.87 kg/m^2

Conveyor length: 110 m

Coefficient of friction (rollers) 0.1

W = Total weight = 7.3 + 1.87 = 9.17 kg/m^2

L = Conveyor length = 110 m

w = Band weight = 7.3 kg/m^2

f_r = Rollers = 0.1

$T = (9.17 \times 110 \times 0.1 + 7.3 \times 110 \times 0.1)$ kg/m of band width

$T = (100.87 + 80.3)$

$T = 181.17$ kg/m of band width

For band width of 1.5 m

Tension = 271.75 kg

11.5 CALCULATION OF TORQUE REQUIRED FOR THE CONVEYOR DRIVE

We need to take into consideration the efficiency of our drive arrangement. We use standard factors/service data for each section of the drive arrangement.

- The toothed belt drive between the motor and the gearbox is 95% efficient giving a factor of 1.05
- The gearbox has a service factor given in the data sheet of 1.04
- For the chain drives, the factor is determined by the size of the chain sprockets (the smallest size in each case) and is shown in the supplier's (Cross and Morse) manual
- First stage (18-tooth sprocket): 1.45
- Second stage (17-tooth sprocket): 1.55

We therefore have a factor for our drive design of:
1.05 × 1.04 × 1.45 × 1.55 = **2.454**

11.5.1 Drive Torque

Torque = Tension × 0.5 × drum diameter

For a drum 1.1 m diameter

Torque = 271.75 × 0.5 × 1.1 = 149.46 kg force

Torque = 1466 Nm

1 kg force = 9.807 Nm

11.6 CALCULATION OF ELECTRIC MOTOR POWER

Power (kW) = Torque (Nm) × Speed (rpm)/9554

11.6.1 Example

The torque requirement is **1466 Nm**

The drive efficiency factor is **2.454**

Oven baking time 3.0 minutes on 100 m long oven gives band speed of 33.33 m/minute.

Drum diameter 1.1 m with drum circumference of 3.456 m

Drum speed = 33.33/3.456= **9.64 rpm**

Power (kW) = 1466 × (9.64/9554) × 2.454

Power (kW) = 3.63

Note: 9554 is a constant which changes rpm to mm/second using the radius component of the torque. It is calculated as (60 × 1000)/2 × 3.142 (Pi) = 9554 (Figure 11.28).

Figure 11.28 Delivery end and transfer to cooling conveyors.

BIBLIOGRAPHY

Cross and Morse, 2015. <www.cross-morse.co.uk>.
Erskine, 2015. <www.dalepowersolutions.com/ups.
Festo, 2015. <www.festo.com>.
Martonair, 2015. <www.martonair.co.uk; www.norgren.com>.
Rexroth Bosch, 2014. <www.boschrexroth.com>.
Synatel, 2015. <www.synatel.co.uk>.

CHAPTER 12

Process Control Systems

Contents

12.1 TEMPERATURE CONTROL

12.1.1 Direct Gas-Fired Ovens

Direct gas-fired (DGF) ovens have the most complete system of temperature control. The system can provide separate top and bottom control of heat input from the burners and lateral control across the width of the oven with multilane distributor burners.

12.1.2 Indirect Radiant Ovens

Indirect ovens have a single burner in each zone, so the temperature is not variable top to bottom or across the width of the oven. Baking control is by varying the airflow to the top and bottom radiant tubes. Lateral control is by means of adjustable restriction plates on the inlet to the radiant tubes (this is set at commissioning to provide an even bake across the width of the oven).

12.1.3 Convection Ovens

Similar to indirect radiant ovens, convection ovens have only one burner in each zone. The temperature of the airflow to the top and bottom of the oven and across the width of the oven is the same. The airflow to the top/bottom of the baking chamber is varied by damper controls.

12.2 TEMPERATURE MONITORING AND CONTROL

The temperature is controlled separately in each oven zone. The temperature is detected within the baking chamber by 'K' type thermocouples. The thermocouples are connected to the temperature controllers by special screened cable: NiCr/Ni type K (Figure 12.1).

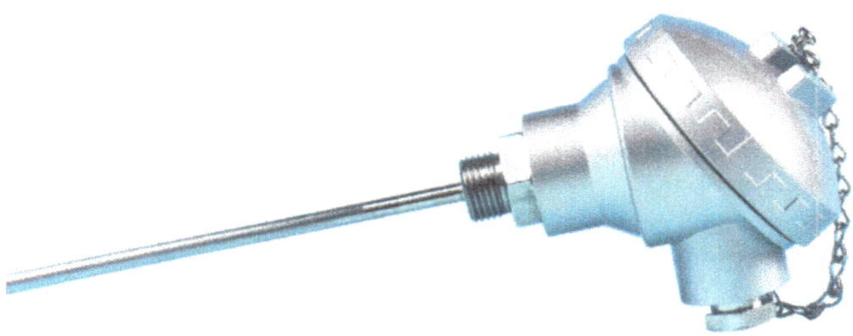

Figure 12.1 'K' type thermocouple with insulated head from TC Direct.

Normally, in a DGF oven, there will be four thermocouples in each zone, two above the band and two below. The thermocouples are connected to temperature controllers on the zone control panel. The controller will average the temperatures from the two thermocouples at the top of the baking chamber and the two thermocouples at the bottom of the oven. In a DGF oven there may be a separate controller and separate system for controlling the top burners and the bottom burners. Alternatively a heat ratio system may be used to adjust top and bottom heat inputs (see Section 12.5).

The position of the thermocouples is important, and the location should not be close to a burner flame, which will give a misleading temperature reading. The location should be adjacent to the back of the burner (opposite side to the flame) and as far from the burners as possible. Locations close to an inspection door or at a zone end are often chosen.

For *indirect radiant or convection ovens*, the control is for a single burner and the oven may have a single or multiple thermocouples from which the temperatures are averaged by the controller. Thermocouples would normally be located in the baking chamber on each side of the heater module.

12.3 TEMPERATURE CONTROLLERS

The thermocouples provide temperature data to automatic controllers. The required baking temperature is set at the controller by push button and the controller will display the set temperature and the actual temperature.

In addition, a simple controller is used to detect an over temperature condition which could be unsafe. A thermocouple in the zone flue or heat exchanger will detect an excessive temperature (as set on the controller) and will automatically shut down the burner system by operating the gas solenoid valve if the temperature exceeds the maximum set point (Figure 12.2).

12.4 PID CONTROL

Temperature controllers use PID: proportional-integral-derivative. PID is a control loop feedback system.

The controller monitors the actual temperature compared to the set point and activates the burner control to maintain the set point temperature. By tuning the three parameters of the PID controller algorithm, the system will respond accurately and promptly to variations between the set point and the actual

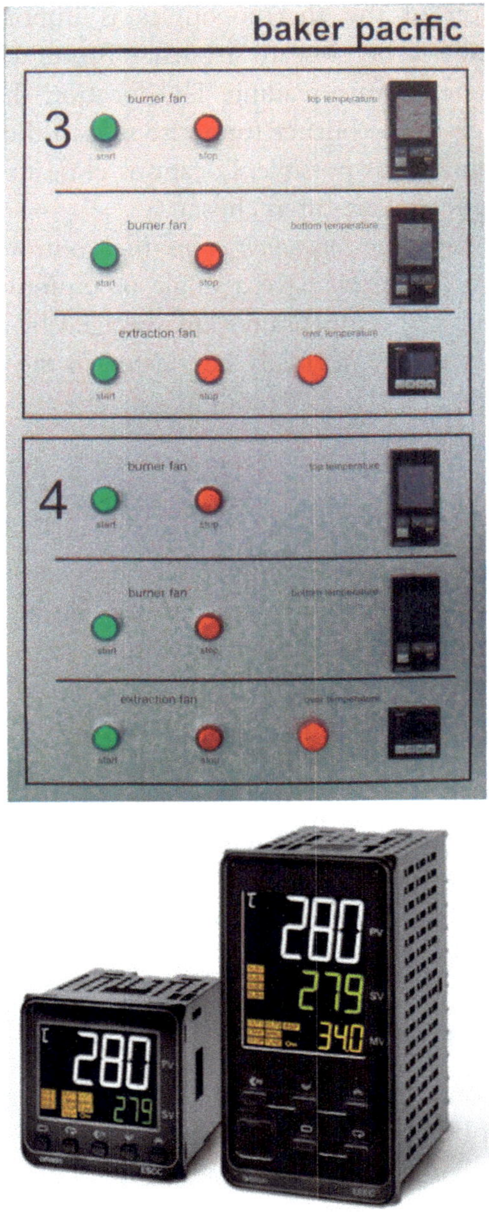

Figure 12.2 Zone control panel for two zones on a direct gas-fired oven with Omron temperature controllers. Each zone has a separate top and bottom controller and a safety override controller.

baking temperatures. When the PID is tuned with the correct values for P, I and D, the system will be responsive and will maintain the baking temperature within ±1°C avoiding overshooting and undershooting.

Proportional term produces an output value proportional to the current error value (the difference between set and actual temperatures). The controller response can be adjusted by the proportional gain constant. A high proportional gain results in a large change in output for a given change in error.

Integral term is proportional to the magnitude and the duration of the error.

Derivative term slows the rate of change of the controller output. Derivative term adjustment can be used to reduce overshoot and improve process-controller stability (Figure 12.3).

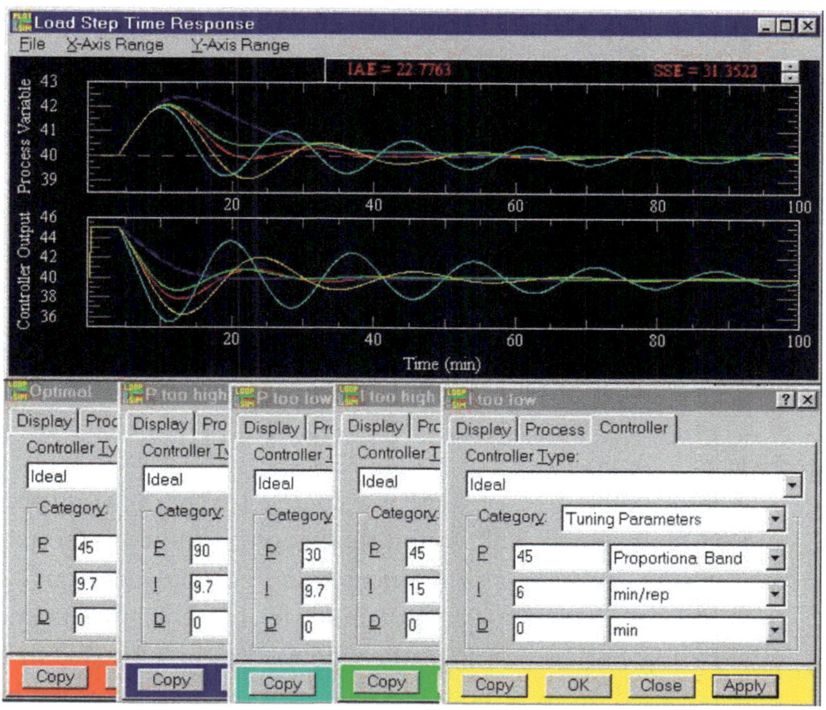

Figure 12.3 PID tuning: the red trace is the optimum. Charts from Metso Expertune www.expertune.com.

12.5 TOP AND BOTTOM TEMPERATURE CONTROL AND HEAT RATIO CONTROL

There are two systems of temperature control used on DGF ovens:

1. Separate top and bottom control

   In this system, the top burners and the bottom burners in each zone have independent temperature controllers and air supply via separate motorised valves (Figure 12.4).

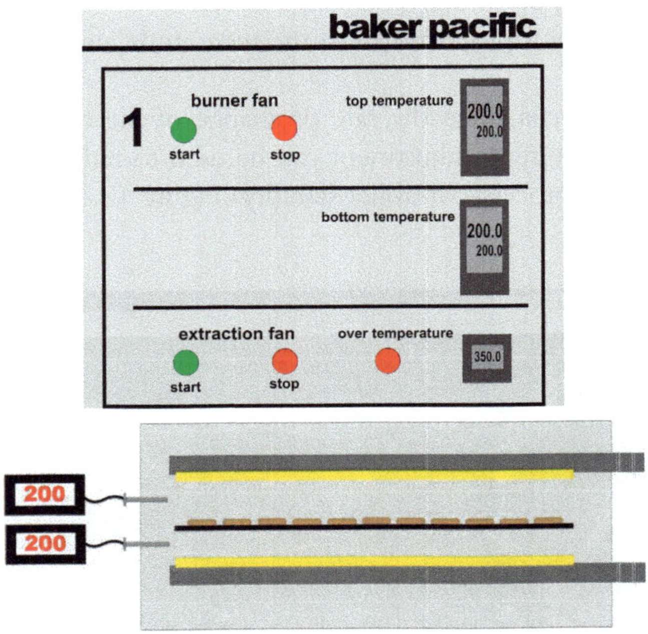

Figure 12.4 Separate top and bottom temperature control.

2. Heat ratio control

   For heat ratio control, there is a single temperature controller and set temperature. The system then controls the top and bottom burners according to a ratio setting. The operator can set the ratio, for example, 60% heat input by bottom burners and 40% by top burners, or any other required ratio. This system is now not used on most new ovens and has been super-ceded by the separate top and bottom control systems (Figure 12.5).

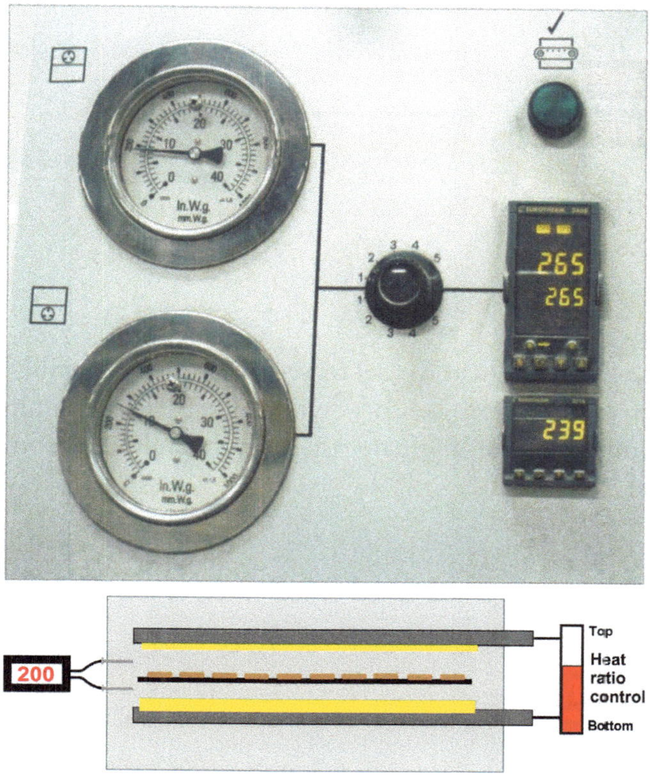

Figure 12.5 Heat ratio control panel.

12.5.1 Separate Top and Bottom Temperature Control

It should be noted that a change in setting to either the top or bottom burners will affect both the top and bottom systems. The oven has one baking chamber, but two separate temperature control systems which interact.

For example, if a change is made to the bottom temperature setting, the bottom burners will increase or decrease the heat input. If the bottom heat is increased, more hot air will rise in the baking chamber and the top temperature sensors will detect oven temperature. The controller will then automatically reduce the heat input from the top burners (Figure 12.6).

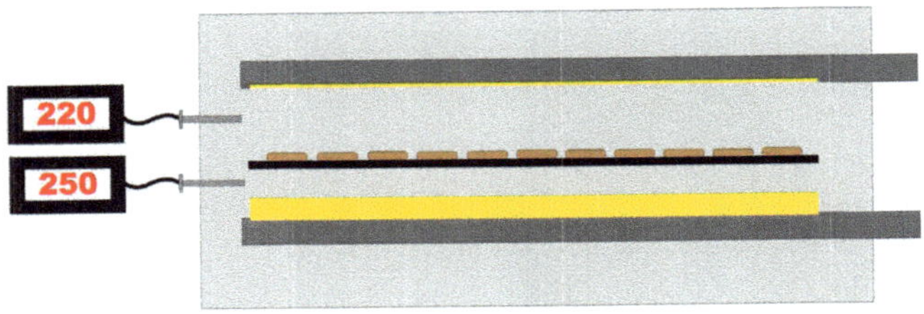

Figure 12.6

If the bottom heat is decreased, less hot air will rise and the top temperature sensors will detect under temperature. The controller will increase the top burners to compensate (Figure 12.7).

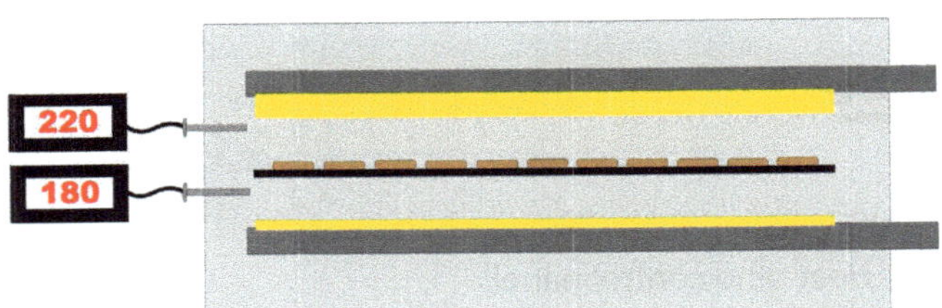

Figure 12.7

12.6 BAKING TIME

The baking time is simply altered by adjusting the conveyor speed. The motor is an AC motor with inverter control. The speed range will depend on the range of products to be made: for example, a cracker oven may have a baking time range of 2.5–7.5 minutes, a cookie oven from 4.0 to 12.0 minutes (Figure 12.8).

Figure 12.8 The PAX Lite Process Time indicator from Red Lion Controls is located on the main control panel.

12.7 HUMIDITY

Control of the humidity in the baking chamber is by extraction of the moisture through a fan and flue. Each zone has a series of extraction duct inlets (cross ducts) at the top of the baking chamber which connect to an extraction duct along the length of the zone. The extraction duct conveys the wet air to the extraction fan, which exhausts it through the flue.

During the first part of the baking process, the baking chamber has high humidity to enable the biscuit structure to form without the outer skin of the dough pieces drying. After the structure of the biscuits is formed, the baking process requires moisture to be evaporated from the dough pieces. The extraction system exhausts the moisture from the baking chamber in order to reduce the moisture content of the biscuits to the required level.

If a DGF oven is burning liquid petroleum gas (LPG) fuel, extraction ducts should be installed at the top and bottom of the baking chamber. LPG is heavier than air, and unburnt gas can lie at the bottom of the baking chamber and cause an explosion if ignited (Figures 12.9–12.12).

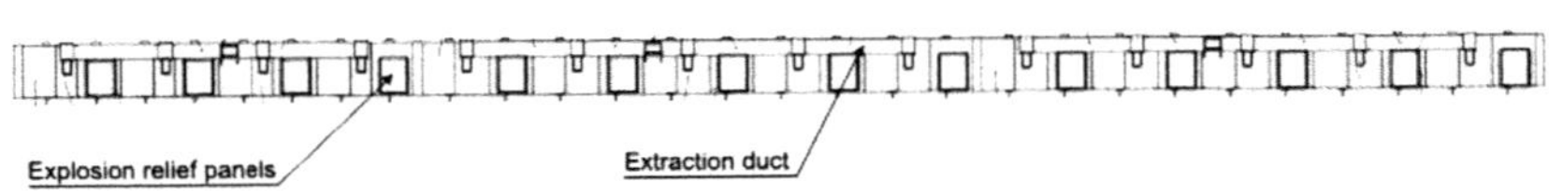

Figure 12.9 Plan of the top of the baking chamber showing position of the extraction duct.

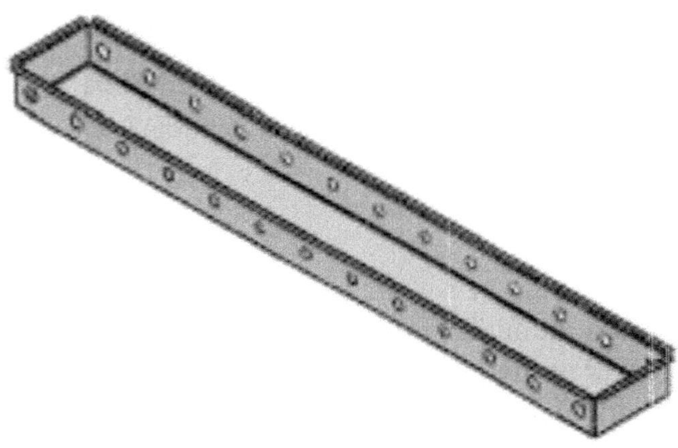

Figure 12.10 Extraction cross duct inside baking chamber.

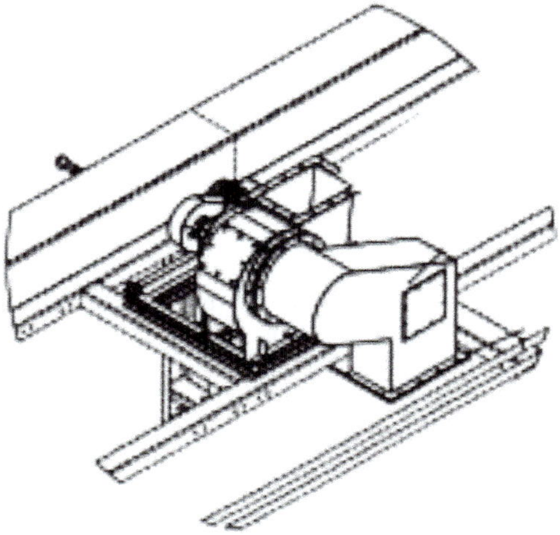

Figure 12.11 Extraction duct and fan on top of the oven.

Figure 12.12 Halifax 12P multivane extraction fan with manual damper control.

The amount of air extracted from the baking chamber is controlled by a damper in the fan inlet duct. This may be controlled manually or by servo motor.

12.8 COLOUR CONTROL

The colour is mainly controlled in the final zones of the oven. The biscuit colour develops rapidly when the surface of the dough pieces is dry and at a temperature over 150°C.

A typical problem may be poor lateral control (across the width of the band). This may apply to both colour and moisture content. The first check for problems of lateral control should be at the forming machines. If the dough piece weights vary across the width of the band, the colour and moisture content after baking will also vary.

Ovens may have systems of lateral baking control. These may be summarised as:

- DGF ovens: Multilane burners
- Indirect radiant ovens: Control of airflow to the radiant tubes

DGF ovens: multilane burners allow the burner flame to be adjusted in several lanes across the width of the oven. The burners may have three or five lanes. By adjusting the flame in the centre and/or at each side, the heat input can be balanced and an even colour achieved. In order for the system to be effective, at least 15% of the burners should be multilane (Figure 12.13).

Indirect radiant ovens bake by radiation from tubes carrying the hot air/gases from the burner. The hot gases enter the radiant tubes from distribution ducts across the width of the oven. In these ducts there are sliding plates which carry adjustable restrictors which will reduce the hot gases entering each radiant tube. These restrictor plates are adjusted during commissioning and ensure well-balanced lateral heating (Figure 12.14).

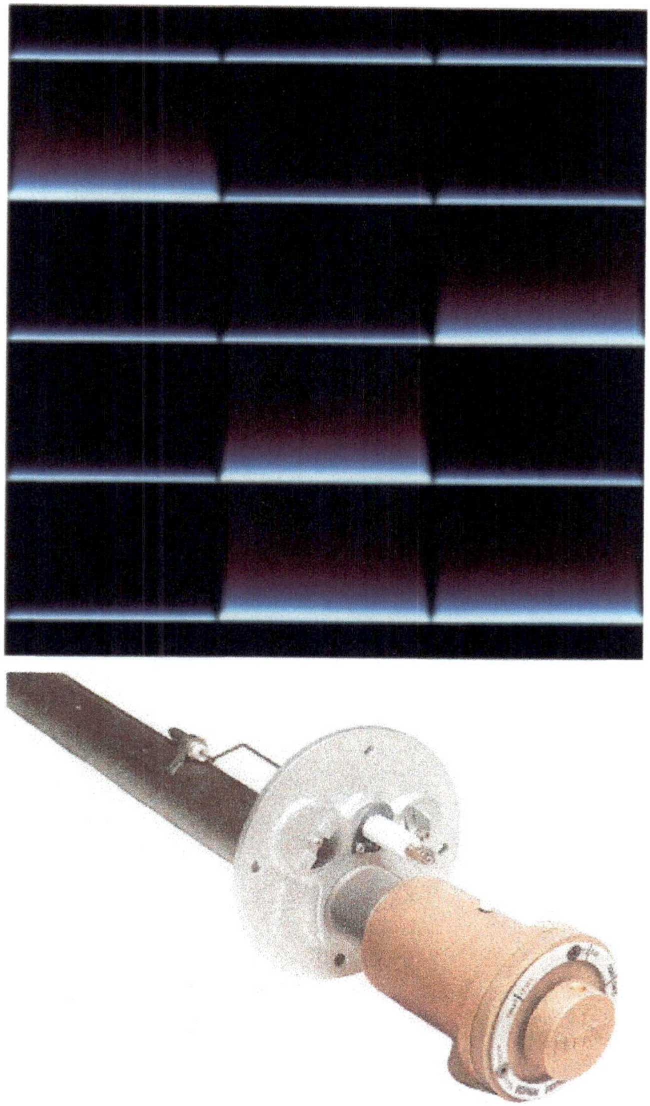

Figure 12.13 Flynn three-lane distributor burner.

Figure 12.14 Indirect radiant oven: distributor ducts with restrictor plates for the radiant tubes.

12.9 COLOUR MEASUREMENT

Usually examples of the correct biscuit colour will be held on a frame at the end of the oven for comparison with the baked biscuits being made. The oven operator may visually check the bake against samples of the correct colour range. For the laboratory a spectrophotometer will be used (Figure 12.15).

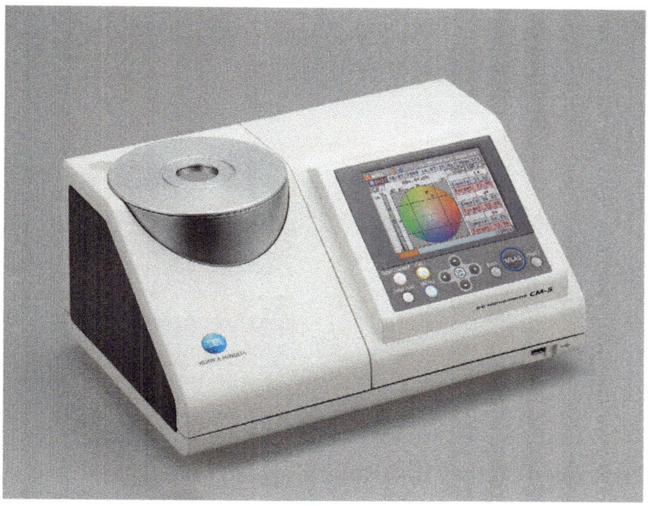

Figure 12.15 Konica Minolta CM-5 spectrophotometer.

The spectrophotometer allows the technician to set up measurements of reflectance and transmittance and a judgement of pass/fail based on a set of tolerances. Data can be shown numerically or on spectral graphs.

12.10 PLC CONTROL

Programmable Logic Controller (PLC) control, graphics with touch screen Human Machine Interface (HMI) or keyboard can provide all the baking control functions and displays. These may include all monitoring and control functions as described above and diagnostics, fault finding and alarm/warning alerts. The equipment control for dampers, valves, etc., will be motorised using servo motors and motor speeds controlled by inverters. The control system can be recipe driven, providing the correct settings according to the recipe selected. The control/monitoring is presented to the oven operator by an HMI, for example, Allen Bradley PanelView (Figure 12.16).

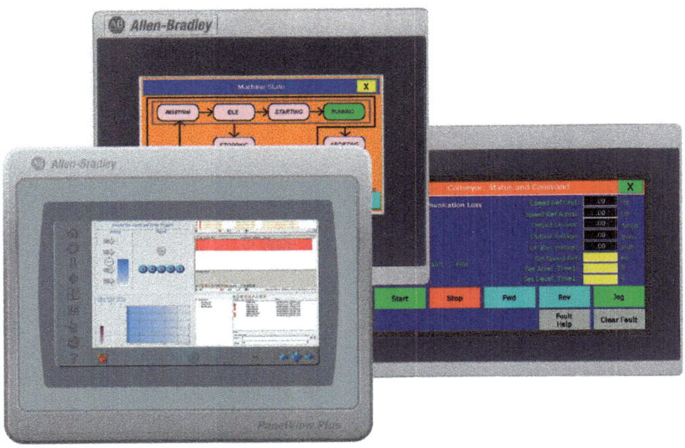

Figure 12.16 Allen Bradley HMI Panel View Plus 7.

The Allen Bradley operator terminals are ideal for baking oven applications that require monitoring, controlling and displaying information in dynamic ways, where operators must quickly understand machine status and make better decisions.

In addition to the control, display and alarm functions, the system may provide historical data trending to enable the technicians to better manage recurring faults, downtime, product variations in colour, moisture content and other operational factors.

The oven PLC/HMI system can be linked to a SCADA (supervisory control and data acquisition) system which may include multiple production lines and factory sites. The SCADA system's database and software can provide trending, diagnostic data and management information such as maintenance schedules and reports, expert system trouble-shooting, comparison data of oven performance.

A leading software is Wonderware InTouch from the Schneider Electric Group.

BIBLIOGRAPHY

Antunes pressure switches, 2015. <www.ajantunes.com>.
Allen Bradley, 2015. <www.rockwellautomation.com>.
Flynn Burner Corp, 2015. <www.flynnburner.com>.
Konica Minolta, <www.konicaminolta.com>.
Omron, 2015. <www.ia.omron.com>.
Metso Expertune, 2015. <www.expertune.com>.
Red Lion Controls, 2015. <www.redlion.net>.
Synatel, 2015. <www.synatel.co.uk>.
Weishaupt, 2015. <www.weishaupt-corp.com>.
Wonder Ware In Touch, 2015. <global.wonderware.com>.

CHAPTER 13

Oven Safety Monitoring and Alarm

Contents

13.1 OVEN BAND SAFETY SYSTEMS

13.1.1 Oven Band Tracking

It is essential that the oven band runs in the centre of the baking chamber, with the minimum movement from side to side. If the band deviates excessively the band edges can contact side guide rollers or cast iron side guides. More deviation will result in the band contacting oven parts which will damage the edge of the band.

If the band wanders to one side, there are several possible causes:

1. Uneven band loading (dough pieces missing or not consistently panned evenly across the width of the oven)
2. Uneven heat application by direct gas-fired (DGF) burners or convection system
3. Poor oven band joint with tight side and slack side in the band circuit
4. Oven band support rollers or skids neither level nor at right angles to direction of band travel
5. Oven terminal drums not correctly aligned
6. Lack of band tension or uneven tension
7. Oven tension end drum not moving freely and evenly on the slides
8. Poor quality wire mesh band with uneven mesh.

13.1.2 Detection of the Oven Band Position

The oven band will be monitored for excess movement using limit switches at each end of the oven (Figure 13.1).

Figure 13.1 Oven band tracking system in front of the terminal drum at the delivery end of the oven. Note the side guide rollers to detect the band movement and air cylinders controlling the tracking roll.

If the oven band moves 10 mm to one side, the vertical roller will be pushed and this will actuate the Euchner limit switch. An alarm will sound and the tracking roller will be automatically moved to correct the tracking and bring the oven band back to a central position.

If the band continues to move, at 25 mm deviation, the trip position, the system will automatically sound the alarm and stop the oven band drive.

In this case the engineer must rapidly assess the problem and take corrective action. Once the band can be restarted on slow speed, an override control may be used to start the drive and run until the tracking system brings the band back to a central position.

Note: Settings for the multiposition limit switch are recommended for the band deviation of 10 mm (alarm warning) and 25 mm (band stops). These settings may be adjusted by the engineer and should be frequently checked.

13.1.3 Oven Band Drive

The oven band is driven by an AC motor with inverter control. In order to start the oven band drive, the band must be central on each terminal drum, and the air supply to the cylinders for tensioning the band must be operating. The band tension is monitored and tension must be present before starting the oven band.

The rotation of the oven band drive is monitored by a Synatel unit mounted on the end of the drum shaft (Figure 13.2).

The oven band drive will stop under certain conditions:

1. Power failure
2. Excessive band movement
3. Failure of band tension system
4. Operation of an emergency stop button.

In the event of a power failure causing the oven band to stop, a system must be immediately employed to run the band and empty the oven of biscuits. If the biscuits are not rapidly conveyed out of the oven, they will overheat and catch fire. The fire may damage the oven band and structure.

Figure 13.2 Synatel 'Wirligig' speed monitor.

When the band stops the solenoid valve for the main gas supply will be automatically closed, and all burners will be extinguished and the fans will stop. The fault must be rapidly found by an engineer. The band can be restarted by using an override pushbutton and the band run slowly to empty the oven of biscuits, until the fault is corrected, and the oven can be run normally under automatic control.

13.1.4 Emergency-Stops

Emergency stop (E-stop) buttons, clearly identified, should be positioned at each end of the oven and on the main control panels. If an E-stop button is actuated, the drive, gas supply and burners will shut down (Figure 13.3).

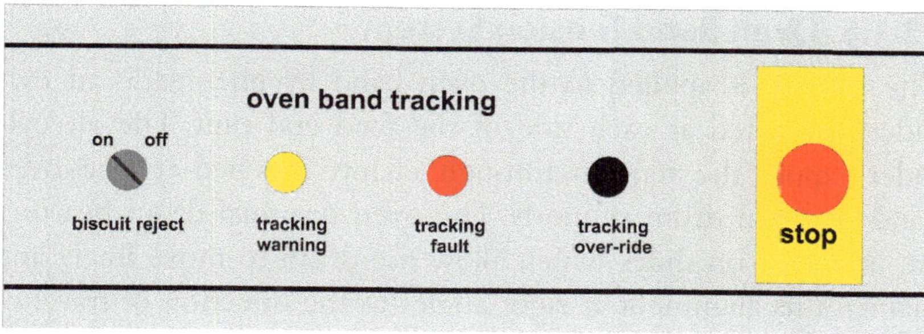

Figure 13.3 Control panel at the oven delivery end with tracking warning lights emergency stop and override button to restart the oven band.

13.1.5 Emergency Oven Band Drive

An automatic system is usually employed to drive the oven band in the event of a power failure. This may be an emergency generator in the factory to provide power immediately to the oven drive or a battery-operated DC motor. For small ovens, for example, 20–30 m long, a manual oven band wind out system may be employed.

If a DC motor is used with battery power, during the emergency wind out (approximately 12 minutes), all safety systems will operate, but the fans are not powered. With a generator, three-phase UPS, all systems will operate, including the fans (Figure 13.4).

Figure 13.4 Oven drive: main AC motor at top right emergency DC motor at bottom right.

13.1.6 Oven Band Tension System

The tension is applied to the oven band by pneumatic air cylinders mounted at each side of the feed end unit. The air cylinders apply the tension through chains attached to the oven band terminal drum supports. The oven terminal drum bearings are mounted on slides which allow the drum to move freely and maintain its alignment at right angles to the direction of travel of the band (Figure 13.5).

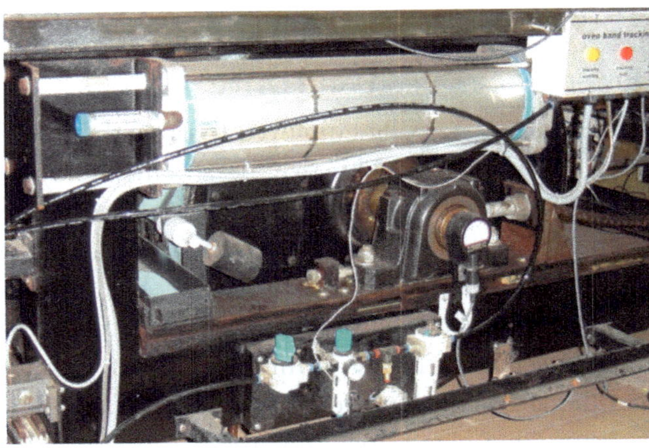

Figure 13.5 Pneumatic cylinder with air supply to maintain band tension. The terminal drum bearings are mounted on linear slide bearings.

A compressed air supply provides air to the two pneumatic cylinders. The regulator that operates the cylinders is set at 2.7–3.4 bar (40–50 psi) on the regulator dial. At commissioning the band tension is set by measuring the amount of sag of the band between two supporting rollers. This amount varies with each band type and may be specified by the band supplier. The air pressure is adjusted to achieve the specified sag in the band.

The air supply to the cylinders is applied via a reservoir so that the pressure is maintained in the event of a failure of the air supply. The system is monitored by a pressure switch in the air supply, and a fault will cause an alarm and warning light on the main panel to show (Figure 13.6).

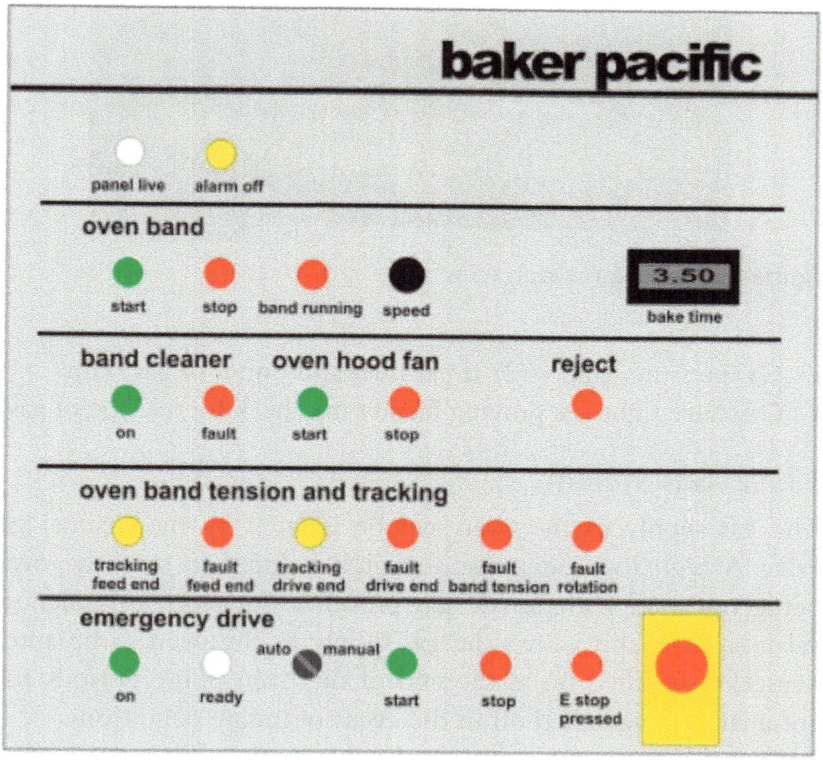

Figure 13.6 Main oven control panel showing warning lights for oven band rotation, tension and tracking.

13.2 OVEN BURNERS AND GAS SYSTEM

The safe operation of the oven heating system is ensured by the following equipment:

13.2.1 Gas Trains

The gas train should include the following equipment to ensure safe operation of the gas feed to the burners (Figure 13.7):

- Manual shutoff valve
- Two automatic solenoid shutoff valves for safety system
- Gas filter
- Zero pressure gas governor
- Gas pressure high/low detection

Figure 13.7 Main gas supply train.

- Gas pressure gauges (2) at gas inlet and outlet of gas train
- Gas valve tightness proving facility (to check for leakage of gas).

13.2.2 Gas System

The gas supply to the oven can be isolated by the manual gas valve. A high/low gas pressure switch is fitted to stop the oven being operated with unsafe gas pressures. An electrical solenoid valve is fitted; this allows the gas supply to the oven to be automatically switched on at the end of the oven purge periods and automatically switched off in the event of unsafe conditions.

The gas leakage system checks that both of the main safety shutoff valves are operating correctly and able to completely stop the gas supply.

13.2.3 Main Gas Solenoid Valves

These valves are situated in the gas train to the oven. Their function is to open on completion of the oven purge and shut off immediately a fault is detected.

13.2.4 High-/Low-Pressure Switch

The gas high-/low-pressure switch is situated in the main gas supply to the oven. Its function is to shut off the burners if the gas pressure supply falls below 12 mbar or exceeds 20 mbar.

13.2.5 Gas Leakage

This system checks that both of the main safety shutoff valves are operating correctly, and there are no leakages in the gas train.

13.2.6 Zero Gas Governor

The zero gas governor is normally situated in the gas train. Its function is to provide an even low (3 mbar) gas pressure to the burners in each zone. An alternative system employs solenoid valves at each direct fired burner, which also function as the zero gas governors.

13.2.7 Purge System (DGF and Indirect Fired Ovens)

The extraction fans will operate with the dampers open to purge the baking chamber. The purge cycle is 6 minutes in duration, and this is controlled by a preset timer. During the purge cycle the volume of air in the baking chamber is exhausted and replaced by fresh air five times.

The duration of the purge cycle and the volume of air expelled from the baking chamber are designed to comply with FM Global insurance standards.

The start of the purge cycle and completion is shown by indicator lamps on the main control panel. Following the completion of the purge cycle the burners can be safely ignited (Figure 13.8).

13.2.8 Over Temperature Safety

In addition to the baking temperature controllers a simple controller is used to detect an over temperature condition which could be unsafe. A thermocouple in the zone flue will detect an excessive temperature (as set on the controller), and the safety system will automatically shut down the burners by closing the gas solenoid valves. An alarm will sound.

13.2.9 Manual Control of Top and Bottom Burners

In special circumstances, the engineer may operate the burners by manual control. Manual control may be selected on the temperature controller and the 'Up' and 'Down' push buttons used to achieve the required air pressure to the burners.

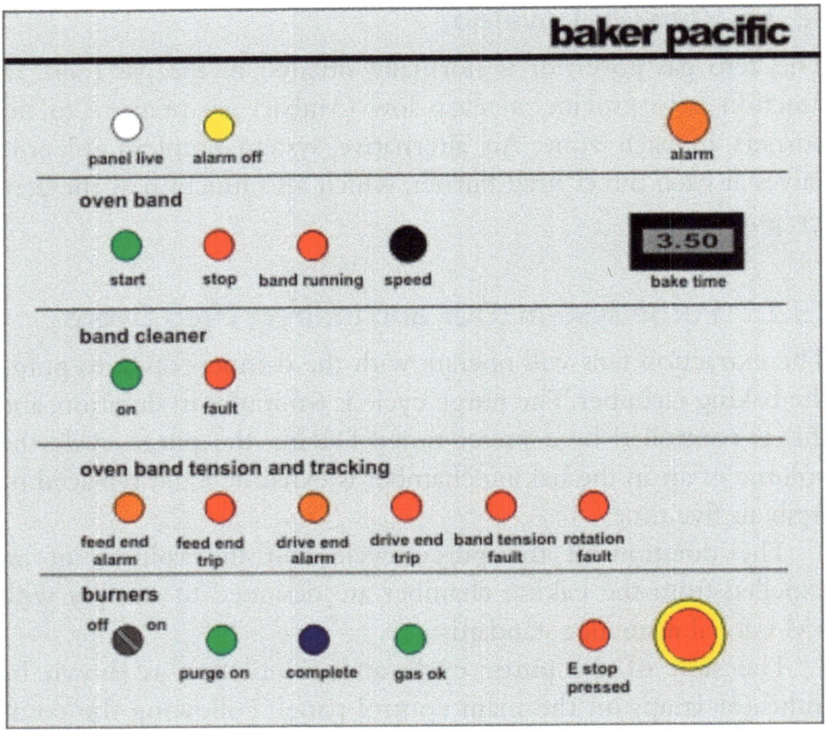

Figure 13.8 Main oven control panel.

13.2.10 Air System

The combustion air is provided by separate fans for the top and bottom burners. The fans have pressure switches to monitor operation.

13.2.11 Burner Flame Supervision

The Flynn Ignition Control Unit provides for automatic ignition and safe supervision of the gas burner.

The DGF burners have a single electrode for ignition and flame detection. These functions are controlled at each burner by an Ignition Control Unit. Ignition is achieved from a spark directed across the burner strip at the electrode from a high-tension transformer in the flame ignition unit.

The start-up sequence is initiated by a switch on the unit or by a switch on the zone control panel. The control starts its sequence by checking that the flame probe (electrode) is signalling no flame. The high-voltage spark generator is then energised together with the solenoid valve. The spark and valve will remain energised until the gas is lit or the ignition safety period is completed.

When the gas is lit the flame is detected by the electrode causing the spark igniter to be de-energised. The solenoid valve will remain open until the electrical supply to the unit is terminated or the flame goes out.

If the burner fails to ignite or the flame subsequently fails, the gas valve will be closed and the control unit will go to lockout. Reset of the lockout followed by a new start-up sequence can only be achieved by interrupting the electrical supply to the unit for a minimum of 3 seconds.

In order to light the burners, the following conditions must be met:

1. The switch on the flame supervision unit is 'On'.
2. The 'ignition' selector switch is turned to 'On'.

The burners will now light up in banks. As an additional safety precaution, there is a 10-second delay between each bank of burners lighting.

The Flynn control unit for each burner allows the burner to be switched on or off. The burner flames are ignited and monitored by an electrode.

A lamp indicates that the burner is on and a flame is present. If the flame fails a warning red light will show and the gas supply to the burner will be shut off by the solenoid valve.

13.2.12 Gas/Air Mixture

Each gas/air mixer has a screw adjustment to alter the ratio of gas to air supplied to the burner. The ratio is set by the engineer during commissioning to achieve an efficient combustion and a steady blue flame.

To weaken the mixture, (ie, less gas), turn the regulating screw clockwise. Turn the screw counter-clockwise to make the mixture richer.

Start with a rich mixture, which is indicated by a green- or red-coloured flame and slowly weaken until the base of the flame is blue. The flame should sit on the strip with a bright blue cone.

Burner heat output is controlled by variation of air pressure, and since the gas is maintained at nominally zero pressure, the gas/air ratio will remain constant and ensure a completely combustible mixture at all air pressures.

13.2.13 Indirect Fired Ovens: Weishaupt Burners

The Weishaupt burner is equipped with a combustion manager and gas pressure monitoring and shutoff valves.

The burner ignition/shutdown sequences are controlled by a dedicated control unit. Flame failure, gas pressure, air pressure and fan pressure are monitored by the controller.

13.2.14 Excess Temperature

The zone temperature is monitored by a separate temperature controller. If the automatic temperature control system fails and the temperature in the baking chamber continues to increase over a safe level, the oven will automatically shut down. An excess temperature is preset on the controller and if the temperature exceeds the preset, the gas supply will be shut off and all burners extinguished.

13.3 EXTRACTION, COMBUSTION AIR AND CIRCULATION FANS

The operation of all fans is monitored by pressure switches. In the case of a fault, the system will give an alarm (sound and warning light on the control panel) (Figure 13.9).

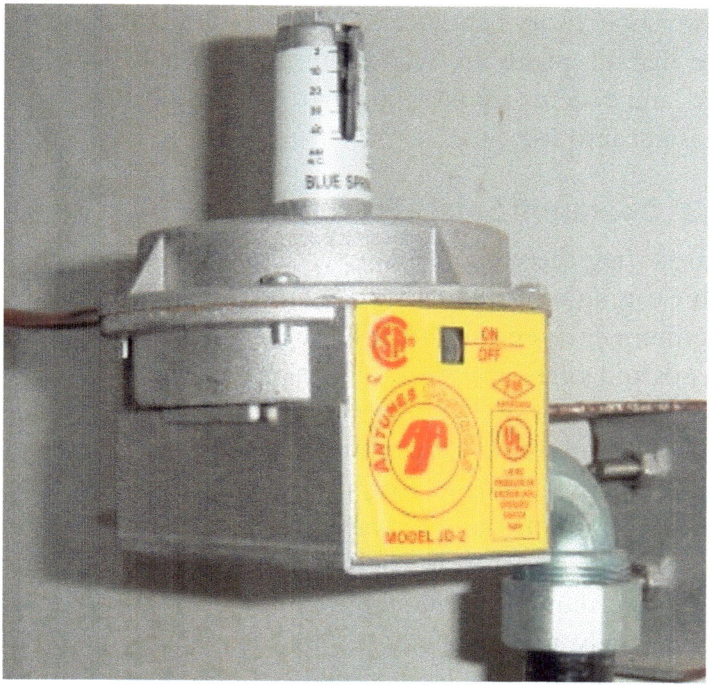

Figure 13.9 Antunes pressure switch for detection of fan operation.

13.3.1 Damper Controls

For burner ignition and the purge cycle, the extraction dampers must be fully open to ensure complete evacuation of any gas in the baking chamber. The damper open position is detected by a limit switch.

13.4 OVEN: GENERAL SAFETY EQUIPMENT AND INSTRUCTIONS

13.4.1 Guards and Safety Devices

Suitable guards and safety devices are incorporated in the oven design following the best industry standards. All safety guards and systems must be fully operational. Users are instructed not to override electrical safety systems or remove guards.

The supervisory engineer installing the equipment must ensure that all guards are secured in position and that the safety devices and systems are set to operate correctly before the equipment is put into use.

13.4.2 Operation Safety Precautions

Before starting the equipment, especially after a shutdown during which cleaning or maintenance activities have taken place, inspect the equipment to ensure that it has been correctly assembled and all guards and safety systems are in place.

The equipment should always be run within the specified limits for speeds, pressures, temperatures, etc.

The equipment should always be attended while in operation, observe indicator lights and warnings that are displayed on the oven panels. Take appropriate action for any alarms and warnings.

The oven requires periodic preventive maintenance to keep it operating safely and efficiently.

Any operational problems should be investigated immediately and the necessary action taken.

For equipment with PLC control, do not attempt to modify the program unless in accordance with the supplier's instructions.

13.4.3 Cleaning and Maintenance Safety Precautions

The user is responsible for ensuring that safeties are in place for isolation of the equipment, before any cleaning or maintenance tasks are commenced. All power, air and fuel sources should be locked out.

Always ensure that the oven has cooled sufficiently before attempting any cleaning or maintenance.

After any maintenance, carefully check that all screws, bolts, nuts and other fixings are securely replaced.

Some cleaning and maintenance operations involve the use of compressed air. Use compressed air carefully.

13.4.4 Commissioning Safety Precautions

Before starting equipment the moving parts must be checked for freedom of movement by rotating by hand where possible. Where necessary, direction of rotation of drives must be confirmed before they are put into operation.

Correct lubrication of the equipment must be checked before it is operated.

Before start-up the electrical equipment must be tested to detect any problems which may have occurred due to dampness or damage caused during transport or storage.

13.4.5 Running the Equipment Without Safety Systems

In normal circumstances the equipment should not be operated unless all safety systems are fully operational. However on very rare occasions for specific tasks, it may be necessary to run the equipment with some guards or safety devices removed. Some electrical maintenance activities may require that the power is turned on.

Note: The user is responsible for implementing safe working practice and procedures for this type of situation. The person who is to do the work must be an employee that the user has authorised and who has received training to be able to carry out the task safely and effectively.

13.4.6 Protection of Employees

Protective equipment should be used as required by the task:

- Hardhats
- Safety glasses
- Ear protection
- Gloves
- Overalls
- Dust masks
- Fume and dust extraction.

13.4.7 Emergency Shutdown

To stop the equipment in an emergency, press one of the red E-stop push buttons located at each end of the oven and on the main control panels.

When an E-stop is pressed the oven band will stop, the fans will stop and the burner system will shut down. The safety problem should be immediately addressed and solved before restarting the oven.

BIBLIOGRAPHY

Baker Perkins Ltd. Oven operation and maintenance manuals. <www.bakerperkins.com>.

CHAPTER 14

Oven Operation: Direct Gas-Fired Oven

Contents

14.1 STARTING THE DIRECT GAS-FIRED OVEN

14.1.1 Preparation

14.1.1.1 Baking Programme

Allow 30 minutes from start up before baking for the required baking temperatures to be reached – high baking temperatures may take longer.

14.1.1.2 Burner Pattern

Normally all burners will be alight, and the automatic temperature control system will adjust the heat input to maintain the set baking temperatures by modulating the burner flames.

If a very low heat input is required for a product, some burners may not be required and the oven may be operated without these burners. Burners can be closed by an on/off switch on the ignition control unit at each burner. The burners switched off should be evenly distributed throughout the zone.

14.1.1.3 Before Starting the Oven

1. Check that the stripping knife at the oven end is disengaged.
2. Check that the oven feed and delivery ends are clear of obstructions.
3. Ensure that all inspection and access doors are closed.
4. Before starting the oven, ensure that the air supply is turned on for the band tension and band tracking systems.
5. Open the main supply manual gas valve.
6. Switch on the main control panel isolator.
7. Switch on the zone panel isolators (Figure 14.1).

Figure 14.1 Main control panel and zone control panels at the oven end.

14.1.1.4 Zone Control Panels

See Figure 14.2.

At each zone panel

1. Start the extraction fan.
2. Start the burner air fan.
3. Set the temperature controllers to the required temperatures.

14.1.1.5 Main Control Panel

See Figure 14.3.

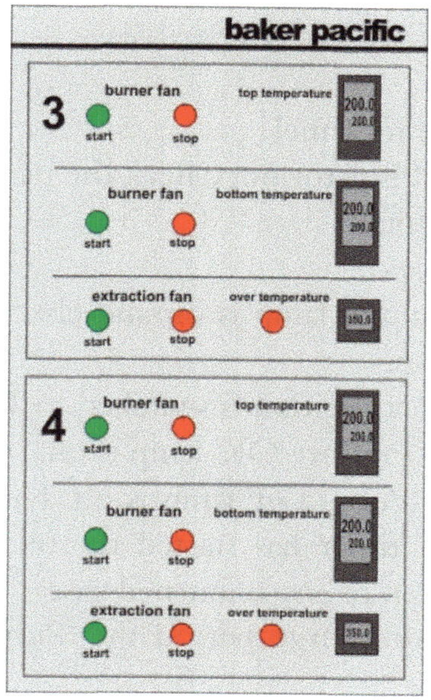

Figure 14.2 Zone control panel.

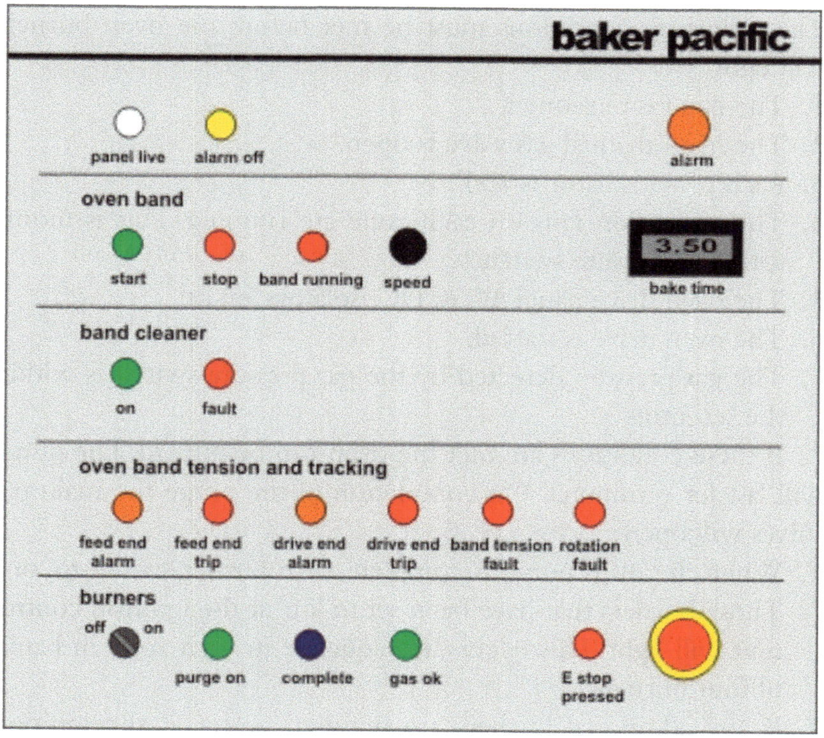

Figure 14.3 Main control panel.

At the main control panel

1. Switch the panel isolator on. Check the panel live lamp is on.
2. Set the baking time.
3. Start the oven band.
4. Check band running lamp is on and check band tension and tracking warnings.
5. Check that burner ignition is switched to 'Off'.
6. Check that 'Gas Pressure OK' lamp is on.
7. Check that the 'Purge On' lamp is on. This indicates that the 6-minute purge timer has started to run to ensure that the fans clear the oven of any unburned gases.
8. At the end of the purge period the 'Purge Complete' lamp will be on.

14.1.2 Lighting the Burners – Direct Gas-Fired Zone

The following conditions must be met before the oven burners will light:

1. The band is tensioned.
2. The main manual gas valve is open.
3. Each panel isolator is 'On'.
4. The extraction fans on each zone are running. This is monitored by pressure switches.
5. The oven drive panel AC & DC isolators are on.
6. The oven drive is started.
7. The gas pressure detected by the gas pressure switch is within the set limits.

If these conditions are met the oven can be purged. The purge will last for 6 minutes. On completion of the purge the main gas valves will open.

1. When the oven purge is complete, turn burner switch to 'on'. Those burners that have been set to 'on' at the ignition control unit will light. This occurs, in sequence in each zone in banks of four burners.
2. If any additional burners are required, select at the ignition control unit.

3. If any burner fails to light, switch the ignition control off, then on again.

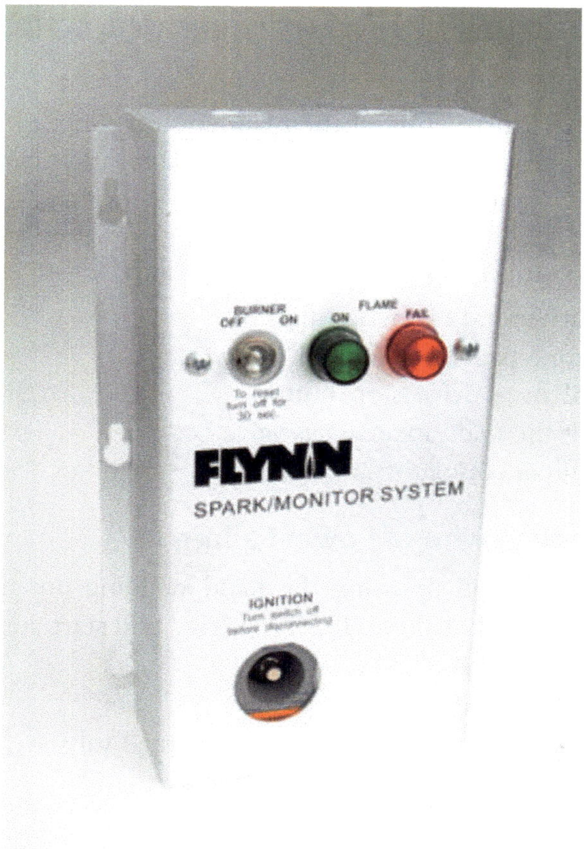

Figure 14.4 Flynn ignition control unit.

14.2 HEATING UP/START OF PRODUCTION

1. Set the extraction dampers to 20% open during the warm up period.
2. Approximately ten minutes before baking commences, set the extraction dampers to their baking settings.

3. Check baking time is correct.
4. Start band cleaner unit.
5. When baking temperatures are correct, start production.

14.3 SHUTTING DOWN THE DIRECT GAS-FIRED OVEN

1. Switch to 'Burner Off' at the main control panel.
2. Set the extraction dampers in each zone to fully open.
3. Keep inspection doors closed.

14.3.1 When Temperature Drops to 100°C

1. Stop burner and extraction fans.
2. Leave zone panel isolators on.
3. Press the oven drive 'stop' button.
4. Turn off the main manual gas valve.
5. Turn off panel isolators.

14.3.2 In the Event of Power Failure

1. Empty the oven by using the hand winding-out mechanism or the DC motor drive. The DC drive will start automatically on power failure.
2. Turn the ignition switches to 'Off'.
3. Turn off the gas supply at the main manual valve.

14.3.3 In an Emergency

1. Press the 'Emergency Stop' button on the main drive panel or at the oven ends.
2. When any danger to personnel is removed, re-set the Emergency Stop and press the 'Main Drive' Start button.

BIBLIOGRAPHY

Baker Perkins Ltd., Oven operation and maintenance manuals. <www.baker-perkins.com>.

CHAPTER 15

Oven Operation: Indirect Radiant Oven

Contents

15.1 OPERATION OF THE INDIRECT RADIANT OVEN

Figure 15.1 Baker Pacific indirect radiant oven.

15.1.1 Preparation

15.1.1.1 Baking Programme

Allow 40 minutes from start-up before baking for the required baking temperatures to be reached; high baking temperatures may take longer.

15.1.1.2 Before Starting the Oven

1. Check that the stripping knife is disengaged.
2. Check that the oven feed and delivery ends are clear of obstructions.
3. Ensure that all inspection and access doors are closed.
4. Before starting the oven, ensure that the air supply is turned on for the band tension and band tracking systems.
5. Open the main supply manual gas valve.
6. Switch on the main control panel isolator.
7. Switch on the zone panel isolators.

15.1.2 Starting the Oven

15.1.2.1 Main Control Panel

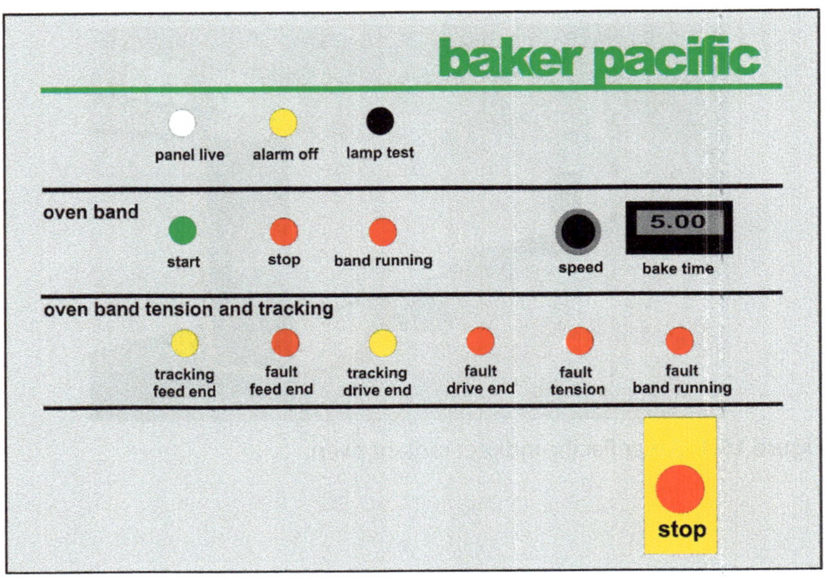

Figure 15.2

1. Switch the panel isolator on. Check panel live lamp is on.
2. Set the baking time.
3. Start the oven band.
4. Check band running lamp is on and check band tension and tracking warnings.
5. Start oven end hood fan (if fitted).

15.1.2.2 Zone Control Panel

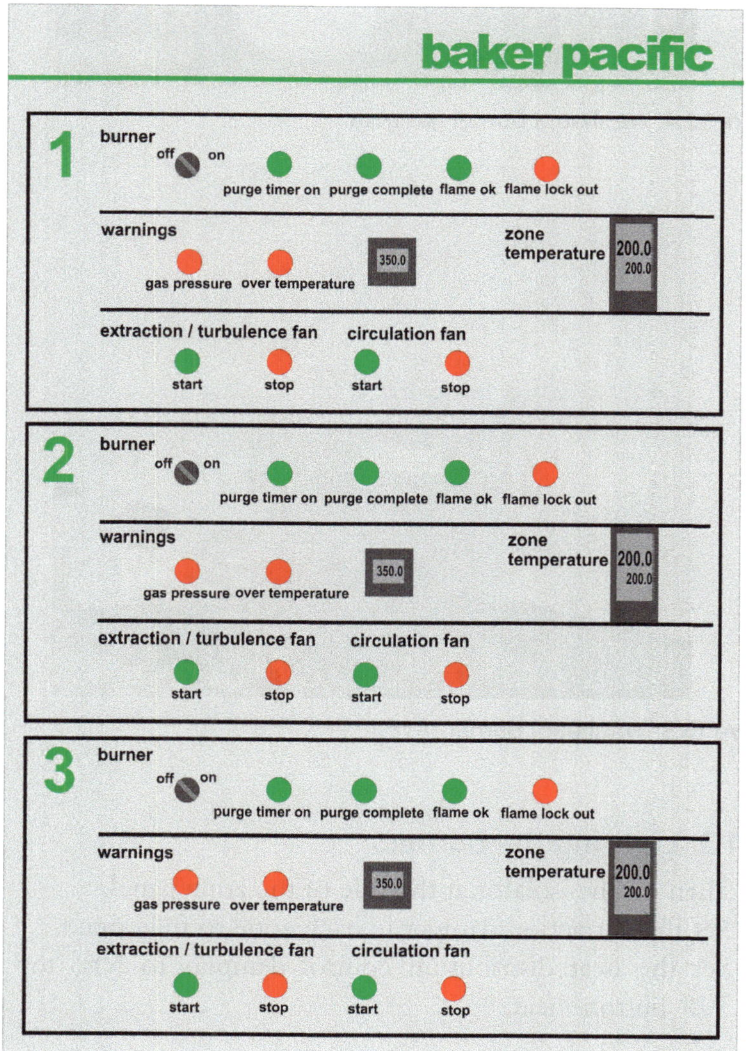

Figure 15.3

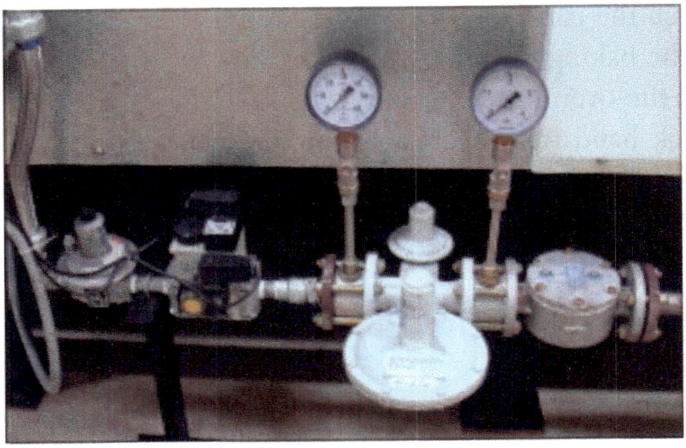

Figure 15.4 Weishaupt burner gas train.

Figure 15.5 Weishaupt burner.

15.1.3 Lighting the Burners

1. Turn on the isolator at the side of the zone panel.
2. Set the extraction damper in each zone to fully open.
3. Set the heat distribution control dampers to 50% top and 50% bottom heat.

4. Switch the extraction and circulation fans on in each zone.
5. Open the main gas supply valve on the gas train.
6. Check gas pressure lamps OK.
7. Turn burner switch to on in each zone.
8. Start purge timer. When the purge is completed 'purge complete', lamp will be on.
9. After the purge cycle (6 minutes) the burners will light. Check 'flame ok' lamps.
10. If the burner develops a fault, the 'Lockout' lamp will come on. Pressing the 'Lockout/Reset' button allows a repeat of the burner lighting procedure.

15.1.4 Heating Up/production

1. Set the extraction dampers to 20% open during the warm-up period.
2. Set the baking temperatures for each zone.
3. Approximately 10 minutes before baking commences, set the extraction dampers to their baking settings.
4. Check baking time is correct.
5. Start band cleaner unit.
6. When baking temperatures are correct, start production.

15.1.5 Damper Controls

Figure 15.6 Damper controls: 1. Zone heating dampers for left and right zone control, 2. Extraction damper, 3. Turbulence controls for top and bottom.

15.1.5.1 Zone Heating Controls

The zone heating dampers control the distribution of heat to the top and bottom of the zone. There are separate dampers for the heat distribution to the left and right of the zone (Fig. 15.7).

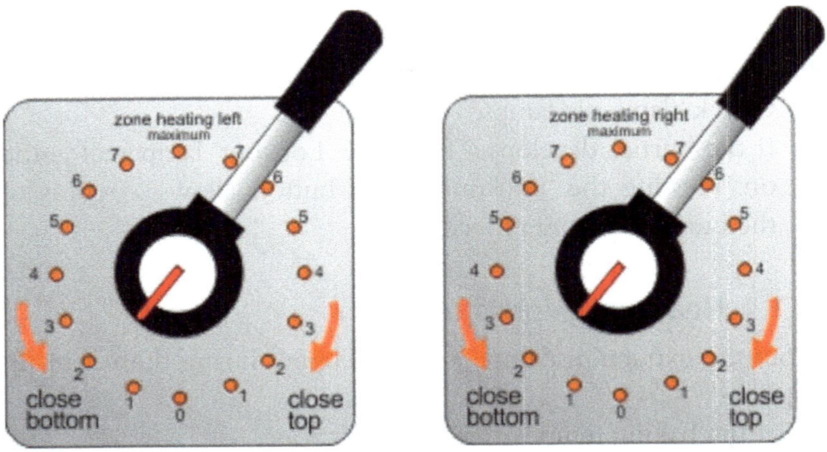

Figure 15.7 Zone heating damper controls.

15.1.5.2 Extraction and Turbulence Dampers

An extraction fan draws the moist air from the baking chamber in each zone. This air can then be exhausted to atmosphere through the flue or it can be directed to the top and bottom turbulence ducts. The damper controls allow the proportion of air extracted from the baking chamber to be divided between the flue and the turbulence ducts (Figs 15.8 and 15.9).

Figure 15.8 Extraction damper control for the amount of air to be exhausted through the extraction chimney.

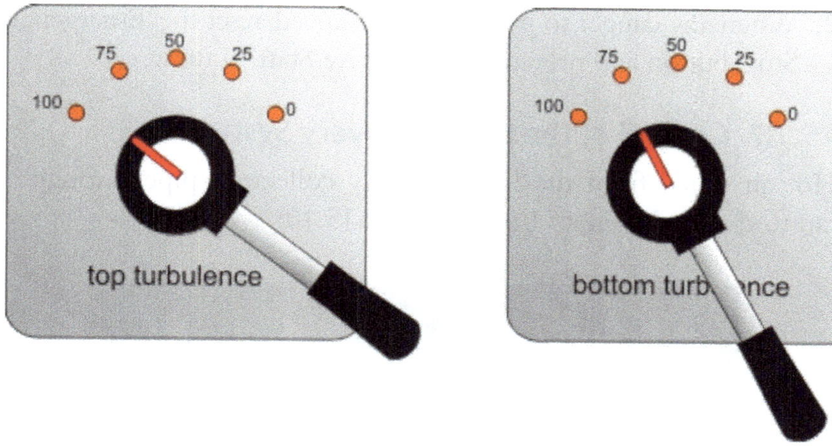

Figure 15.9 Damper controls for top and bottom turbulence.

15.1.6 Shutting Down the Indirect Radiant Oven

1. Switch to 'Burner Off' at the zone control panels.
2. Set the extraction dampers in each zone to fully open.
3. Keep inspection doors closed.

15.1.6.1 When Temperature Drops to 100°C

1. Stop the circulation and extraction fans.
2. Leave zone panel isolators on.
3. Press the oven drive 'stop' button.
4. Turn off the main manual gas valve.
5. Turn off panel isolators.

15.1.6.2 In the Event of Power Failure

1. Empty the oven by using the hand winding-out mechanism or the DC motor drive. The DC drive will start automatically on power failure.
2. Turn the burner switches to 'Off' on the zone panels.
3. Turn off the gas supply at the main manual valve.

15.1.6.3 In an Emergency

1. Press the 'Emergency Stop' button on the main drive panel or at the oven ends.

2. When any danger to personnel is removed, reset the 'Emergency Stop' button and press the 'Main Drive Start' button.

15.1.7 Control of the Heat Recovery System

Hot air is fed from the heat recovery collection pipe through a fan to the final zone of the oven (Fig. 15.10).

Figure 15.10 Baker Pacific indirect radiant oven with heat recovery system.

The amount of hot air/burnt gas taken from each burner flue is controlled by dampers. One damper controls the exhaust flue and one damper controls the connecting pipe from the burner flue to the collection pipe. These dampers are set by the commissioning engineer to allow sufficient quantity of heat for the final zone of the oven (Fig. 15.11).

The final zone of the oven has radiant ducts above and below the oven band. These ducts are divided along their length into three sections (control side, centre and non-control side). The flow of hot air into each section is controlled by a damper. These dampers may be adjusted to ensure the optimum heat balance top to bottom and across the width of the oven (Fig. 15.12).

Figure 15.11 Extraction flue on the left and burner flue on the right with connection to collecting pipe for recovered hot air.

Figure 15.12 Damper controls for the final oven zone heated by recovered hot air. Separate damper controls for top and bottom and control side/centre/non-control side to balance lateral control.

BIBLIOGRAPHY

Baker Perkins Ltd. Oven operation and maintenance manuals. <www.baker-perkins.com>.

CHAPTER 16

Oven Efficiency

Contents

16.1 ENERGY USE

We are concerned not only to bake high-quality biscuits but also to achieve the lowest cost per kilogram of baked product. We need therefore to consider the amount of energy required by the oven. As energy costs increase in almost every country, the efficiency of the oven is of growing importance. In certain countries such as India, fuel is very expensive and represents an important element of the total production cost.

The energy used by the oven is predominantly from gas or oil fuel. Electricity is rarely used for baking now, due to its high cost. In a gas-/oil-fired oven the fuel represents around 95–96%

of the total energy usage and electricity (for powering the drive, fans and other electrical systems) about 4–5%.

The energy input to the oven is used primarily to bake the biscuit, to achieve the structure, reduce the moisture content by evaporation and to colour the biscuit. Each type of biscuit requires a certain amount of energy to achieve a good-quality result. In the example, a typical rotary moulded product that requires 0.2120 kWh (182 kcal) of energy per kilogram of baked product is used.

In addition to the energy required to bake a good product, energy is lost in several ways:

- By extraction of moist air from each oven zone
- By heat loss through the insulation and outer covers of the oven
- By the return circuit of the oven band
- By heat loss from the flues from a heater module or heat exchanger in an indirect fired oven
- By hot air escaping from the oven delivery end (Fig. 16.1).

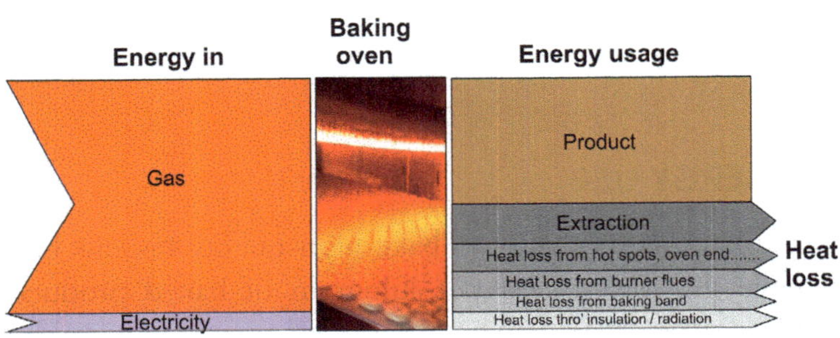

Figure 16.1

In order to minimise the heat loss, (wasted energy), we need to pay attention to the extraction system to achieve the final moisture content required, without excessive waste of heat from the burners, insulate the baking chamber adequately and also insulate the return band, particularly where the band temperature is high, for example when baking crackers.

16.2 EXAMPLE

The following calculations of the energy balance of an oven are taken from an actual installation. Details of the product and the oven are given below, so that different data for other ovens can be substituted to make calculations of energy use accordingly. In the following energy calculations, we will just consider the energy from the gas or oil fuel.

16.2.1 Product

Figure 16.2 Rotary moulded biscuit. Dimensions: 58 mm × 37 mm. Weight: 5.1 g. Baking time: 3.8 minutes.

Indirect Radiant Oven

Baking chamber	1.25 m × 100 m
Zones	8 (7 burners)
Oven band	8.25 kg/m^2
Extraction fans	34 m^3/minute (maximum)
Oven output	3200 kg/hour

Data from independent test results

Total energy used by the oven	0.4043 kWh per kilogram of baked biscuits
Of this, the energy required to bake the product to the required quality	0.2120 kWh per kilogram of baked biscuits
Waste energy	0.1923 kWh per kilogram of baked biscuits

Properties and Conversions

Density of air at 200°C = 0.7461 kg/m^3

Specific heat of air = 1.006 kJ/kgK (at atmospheric pressure)

Specific heat of water vapour: 1.95 kJ/kgK at 200°C

1 kJ = 0.000278 kWh

1 kcal = 0.0011627 kWh.

The total energy requirement to bake the product is 0.2120 kWh per kilogram of baked biscuit.

This gives a total energy usage (gas) to bake the biscuit of:

3200 × 0.2120 = **678.4 kWh/hour**.

This is the energy utilised to bake the biscuit, form the structure, remove the moisture and colour the biscuit.

16.2.2 Latent Heat of Evaporation

An important element in energy use is providing the latent heat of evaporation. In order to evaporate the moisture in the dough (14% by dough weight) to a final moisture content of 3.0%, latent heat is required. The latent heat energy required to evaporate the water from the product is 539 cal/g of water.

Moisture to be removed to reach final moisture content of 3.0%: 0.135 kg per kilogram of biscuits.

Moisture to be removed: 0.135 × 3200 kg = 432 kg/hour.

Latent heat required to evaporate 432 kg of water:

432,000 g × 539 cal = 232848 kcal = 270.73 kWh.

16.2.3 Heat Loss From Extraction System From Baking Chambers

Volume of air extracted from each zone:

34 m^3/minute × 60 minutes × 8 zones = 16320 m^3/hour (maximum).

Estimated average extraction damper setting: 40%.

Estimated volume of air extracted from baking chamber = 6528 m^3/hour.

The air extracted has been heated from ambient temperature over the oven (55°C) to an average baking temperature (200°C). This requires an energy input as follows:

Weight of the air extracted per hour = 6528 × 0.746 kg = 4870 kg.

The energy required to raise the temperature of this air in the oven from 55°C to 200°C

(145°C) is:

145 × 1.006 kJ/kg × 4870 kg = 710387 kJ = **197 kWh/hour**.

Energy required to raise the temperature of the water vapour from 100°C to 200°C:

1.95 × 432 × 100 = 84240 kJ = **23.4 kWh/hour**.

Total heat loss from extraction system per hour = 220 kWh.

16.2.4 Heat Loss From Return Band

Oven band drum centres: 111 m

Band width 1.25 m

Band weight: 8.20 kg/m^2

Specific heat of carbon steel: 0.12 kcal/kg °C

Band temperature at delivery end: 140°C

Return band temperature at feed end: 105°C (estimated temperatures)

Weight of band (on return circuit): 111 × 1.25 × 8.20 = 1138 kg

Temperature loss: 140–105 = 35°C

Heat loss: 0.12 × 1138 × 35 = 4780 kcal (5.56 kWh) per revolution of the band

Bake time: 3.8 minutes

Heat loss per hour: 5.56 kWh × 60/3.8 = 87.8 kWh.

16.2.5 Heat Loss From the Insulation and Outer Covers of the Oven

Oven baking chamber	1.25 m × 100 m
Width over covers	2.3 m
Overall height of covers	2.0 m
Average bake temperature	200°C
Average temp in heater modules	350°C
Average outer side cover temperature	55°C
Average outer top cover temperature	55°C
Mineral wool insulation thickness (s)	200 mm sides and 250 mm top
Mineral wool thermal conductivity (k)	0.04 W/m °C

16.2.6 Heat Loss From Sides and Top of the Oven Through the Insulation

Heat loss = $kA\ dT/d$

Total area of oven sides: 100 m × 2 m × 2 = 400 m^2.

This includes seven heater modules and baking chamber sides.

Area of heater modules on burner side: 13 m^2 × 7 = 91 m^2.

Area of heater modules on nonburner side: 2 m^2 × 7 = 14 m^2.

Total area of heater modules = 105 m^2.

Total area of oven sides (less heater modules) = 295 m^2.

Heat loss from sides of baking chamber sections:

0.04 × 295 m^2 × (200 − 55°C)/0.2 = 8555 W.

Heat loss from heater modules:

0.04 × 105 m^2 × (350− 55°C)/0.2 = 6195 W.

Heat loss from top of oven:

0.04 × 100 m × 2.3 m × (200 − 55°C)/0.25 = 5336 W.

Total heat loss through the insulation of oven sides and top: 20.1 kWh.

16.2.7 Heat Loss From Oven Delivery End

Area of oven delivery end covers and hood: 20 m^2.

Estimated heat loss from radiation at the delivery end: 17.0 kWh.

Estimated heat loss from air escaping from oven end at approximately 180°C:

Estimated volume of air in the baking chamber: 1.7 × 0.8 × 100 = 136 m^3.

Estimated volume of air escaping per min = 136/4.0 = 34 m^3.

Weight of air 34 × 0.746 = 25 kg

25 × 60 × 1.006 × (180–50) = 196170 kJ = 54 kWh.

Estimated heat loss from oven delivery end: 71 kWh.

16.2.8 Heat Loss From Burner Flues

Total energy used: 0.4043 kWh × 3200 kg = 1294 kWh/hour.

7 burners: average energy used per hour per burner = 1294/7 = 185 kWh.

Gas consumption: 185/9.8 kWh/m^3 =18.9 m^3/hour per burner (average).

Gas/air volume required per burner: 18.9 m^3 gas + 310 m^3 air.

Assume that the volume of combustion air drawn in by the burners is exhausted through the burner flues.

Estimated average temperature of flue gases: 200°C.

Gas/air weight at 200°C per burner = 310 × 0.746 kg/m^3 = 231 kg/hour/burner.

Estimated energy required to heat the combustion air: 231 kg × 200 × 1.0 kJ/kg °C = 46200 kJ = **12.8 kWh/hour/burner**.

Total heat lost in burner flues: 7 × 12.8 kWh = 89.6 kWh per hour.

Of this 50% can be used in the heat recovery system.

Combustion process: $CH_4 + 3O_2 = \text{Heat} + 2H_20 + CO_2 + O_2$.

Note 1: For complete combustion 10% excess air is required (this amount can vary considerably depending on the burner and heat exchanger design).

Note 2: Air contains 20.9% oxygen.

From calculations above, the energy consumption of the oven per hour:

For product	678.4 kWh	56%
Heat loss from extraction	220.0	18%
Heat loss from burner flues	89.6	7%
Heat loss from return band	87.8	7%
Estimated heat loss of air from oven end	71.0	6%
Estimated loss from thro' metal, fans, etc.	30.0	3%
Heat loss through insulation	20.1	2%
Heat loss from radiation at oven end	17.0	1%
Total	**1214 kWh**	**100.0%**

Note 1: Estimated accuracy in the assumptions and base data is ±10%.
Note 2: The heat loss can be considerably larger than given depending on the design of the heat exchanger and flue.

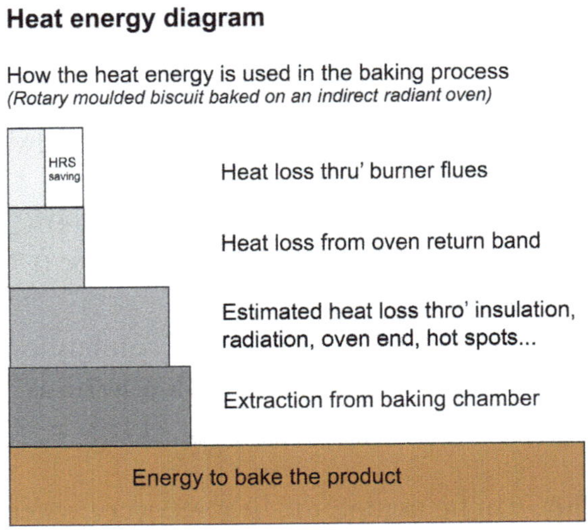

Figure 16.3

- The overall oven efficiency is approximately 56%.
- Of the heat loss through the burner flues, up to 50% can be recovered and used for baking in a *heat recovery system*.
- In this Baker Pacific oven installation (Fig. 16.4), the *heat recovery system* reduced the energy requirement by 15% (as calculated by an independent test).

Figure 16.4 Indirect radiant oven with heat recovery system.

16.3 COMPARISON OF OVEN EFFICIENCY FOR DIFFERENT OVEN TYPES

Table 16.1 Energy usage for various ovens (Based on actual installations)

Product	Oven type	Oven size	kWh/kg of biscuits
Snack cracker	DGF/convection	1.2 m × 90 m	0.477
Rotary moulded	DGF/convection	1.5 m × 100 m	0.441
Rotary moulded	DGF/convection	1.2 m × 60 m	0.430
Rotary moulded	DGF/cyclotherm	1.2 m × 60 m	0.492
Rotary moulded	Indirect radiant + HRS	1.2 m × 100 m	0.404
Rotary moulded	Indirect radiant	1.2 m × 100 m	0.475

16.4 CALCULATIONS FOR THE ENERGY REQUIRED TO BAKE BISCUITS

16.4.1 Rotary Moulded/Hard Sweet/Crackers

The energy requirements are based on the following specific heat data. Specific heat capacity data on ingredients is from www.engineeringtoolbox.com.

Table 16.2 Specific heat of biscuit doughs

Ingredient	Specific heat (kJ/kg K)	Rotary moulded (Glucose)	Hard sweet (Marie)	Cracker (Ritz type)
Flour	1.59	100.0	100.0	100.0
Fat/vegetable oil	1.67	32.1	21.4	11.7
Sugar	1.24	32.5	26.2	10.9
Water	4.20	9.0	20.0	29.5
Total recipe		173.6	167.6	152.1
Specific heat		**1.67**	**1.86**	**2.08**

Specific heat data from The Engineering Tool Box.

Table 16.3 Dry ingredients

Ingredient	Specific heat (kJ/kg K)	Rotary moulded (Glucose)	Hard sweet (Marie)	Cracker (Ritz type)
Flour	1.59	100.0	100.0	100.0
Fat/vegetable oil	1.67	32.1	21.4	11.7
Sugar	1.24	32.5	26.2	10.9
Total		164.6	147.6	122.6
Specific heat		**1.54**	**1.54**	**1.57**

Note: The following calculations for energy requirements are based on the recipes and moisture loss figures given in the Moisture Loss table in the chapter: Baking Process.

16.4.1.1 Rotary Moulded Biscuit

Assumptions:

Production	2000 kg/hour of baked biscuits (2270 kg of dough)
Baking temperature (maximum)	200°C
Ambient temperature	40°C
Water content to be removed	270 kg/hour

Energy is required as follows:

1. The dry ingredients (total 1940 kg) must be raised in temperature from 40°C to 200°C
   1940 kg × 1.54 (specific heat) × 160°C = 478016 kJ = **132.9 kWh/hour.**
2. The water in the dough (270 kg) to be removed must be raised in temperature from 40°C to 100°C
   270 kg × 4.2 (specific heat) × 60°C = 68040 kJ = **18.9 kWh/hour.**
3. Energy must be provided for the latent heat of evaporation
   270,000 g × 539 cal = 145530 kcal = **169.2 kWh/hour.**

4. The water vapour must be raised from 100°C to 200°C

   270 kg × 1.95 (specific heat) × 100°C = 52650 kJ = **14.6 kWh/hour.**

The total energy requirement for baking 2000 kg/h of the rotary moulded biscuits is 335 kWh.

Energy required for rotary moulded biscuit: 0.168 kWh per kilogram of baked biscuits.

16.4.1.2 Hard Sweet Biscuits

Assumptions:

Production	2000 kg/hour of baked biscuits (2460 kg of dough)
Baking temperature (maximum)	220°C
Ambient temperature	40°C
Water content to be removed	460 kg/hour

Energy is required as follows:

1. The dry ingredients (total 1940 kg) must be raised in temperature from 40°C to 220°C

   1940 kg × 1.54 (specific heat) × 180°C = 537768 kJ = **149.5 kWh/hour.**
2. The water in the dough (460 kg) must be raised in temperature from 40°C to 100°C

   460 kg × 4.20 (specific heat) × 60°C = 115920 kJ = **32.2 kWh/hour.**
3. Energy must be provided for the latent heat of evaporation

   460,000 gm × 539 cal = 247940 kcal = **288.3 kWh/hour.**
4. The water vapour must be raised from 100°C to 220°C

   460 kg × 1.95 (specific heat) × 120°C = 107640 kJ = **29.9 KWh/hour.**

The total energy requirement for 2000 kg of the hard sweet biscuits is 500 kWh.

Energy required for hard sweet biscuit: 0.25 kWh per kilogram of baked biscuits.

16.4.1.3 Crackers

Assumptions:

Production	2000 kg/hour of baked crackers (2576 kg of dough)
Baking temperature (maximum)	230°C
Ambient temperature	40°C
Water content to be removed	694 kg/hour

Energy is required as follows:

1. The dry ingredients (total 1950 kg) must be raised in temperature from 40°C to 230°C
   1950 kg × 1.57 (specific heat) × 190°C = 581685 kJ = **161.7 kWh/h.**
2. The water in the dough (630 kgs) must be raised in temperature from 40°C to 100°C
   630 kg × 4.20 (specific heat) × 60°C = 158760 kJ = **44.1 kWh/h.**
3. Energy must be provided for the latent heat of evaporation
   630,000 g × 539 cal = 339570 kcal = **394.8 kWh/h.**
4. The water vapour must be raised from 100°C to 230°C
   630 kg × 1.95 (specific heat) × 130°C = 159705 kJ = **44.4 kWh/h.**

The total energy requirement for 2000 kg of the snack crackers is 645 kWh/hour.

Energy required for baking crackers: 0.32 kWh per kilogram of baked biscuits.

16.5 COMPARING FUEL COSTS

The cost of electricity in most markets has continued to increase. At the same time, natural gas has become more widely available, for example in China and Indonesia. In China, ovens were mainly electric until the 1990s, when town gas and later natural gas became available.

In Indonesia, diesel oil was widely used, with electric heaters to give added direct heat for crackers. In both these countries, gas is now the most economic and the most widely used fuel. There is therefore a strong incentive to convert existing electrically heated oven zones to gas.

The following is a comparison of fuel costs in Indonesia for the first zones of a cracker oven.

Note: Prices of energy vary widely with industry, tariff, date and location; the calculation below is a guide to one such theoretical situation.

16.5.1 Comparison of Fuel Costs: Gas Versus Electricity Versus Oil

IDR, Indonesian Rupiah. USD = IDR 13,000 (May 3, 2015)

Installed capacity		Estimated running capacity (75%)
Oven	540 kW	405 kW (average power)
Preheat	204 kW	204 kW
Total		609 kW

Electricity: cost: IDR 1075/kWh

Running cost per hour: Rp 1075 × 609 = IDR 654675 (USD 50.36)

Natural gas: cost: IDR 820/kWh

Running cost per hour: IDR 820 × 609 = IDR 499380 (USD 38.41)

Diesel Oil: cost: IDR 7500/litre

Calorific value: 11.1 kWh/litre

Oil usage per hour: 609/11.1 = 54.9 litres/hour

Running cost per hour: IDR 7500 × 54.9 = IDR 411750 (USD 31.67).

16.5.2 Payback (Changing Electric Zone to Gas-Fired Zone)

If we assume an investment of USD 120,000 to change the electric zone to a gas-fired zone, the payback is:

Saving per hour: USD 50.36 – 38.41 = USD 11.95

Payback period = 120,000/11.95 hours = 10,042 hours = 1255 × 8-hour shifts

On a three-shift basis, the payback period would be 418 days.

The actual payback period will be shorter as the maintenance cost of the electric oven is very high compared to the gas oven, due to the short life of the electrical heater elements.

Figure 16.5 Gas conversion for an electric zone.

BIBLIOGRAPHY

Alakali, J.S., et al., 2011. Specific Heat Capacity of Palm Crude Oil. Department of Biosource Engineering, McGill University, Canada, Department of Food Science, University of Agriculture, Makurdi, Nigeria.

Armstrong Group, Specific heat – specific gravity. <www.armstrong-intl.com>.

BBC, 2014. Indonesia raises fuel prices by more than 30%. <www.bbc.co.uk>.

Bizee Software Ltd., kW and kWh explained. <www.energylens.com>.

CEIC, 2014. Keeping up with Indonesia's rising energy needs. <www.ceicdata.com>.

Global Business Guide, 2014. Indonesia's subsidies on electric powered down. <www.gbgindonesia.com>.

Hyper Physics, 2015. Specific heat. <www.hyperphysics.phy-astr.gsu.edu>.

Kim, Y.S., et al., 2003. Physical, Chemical and Thermal Characterisation of Wheat Flour Products. Department of Biological and Agriculture Engineering. Kansas State University, Kansas, USA.

Lantau Group, 2014. Gas pricing in Southeast Asia. <www.lantaugroup.com>.

Sugar Engineers, Specific heat capacity. <www.sugartech.co.za>.

Testo, Inc., 2015. Combustion analysis. <www.testo.com>.

The Boiling Point of Water and Solutions, 2015. http://www.chestofbooks.com/food/science/Experimental-Cookery/The-Boiling-Point-of-Water.

The Engineering ToolBox, 2014. Thermal properties for water. Thermal properties for air, food and foodstuff, specific heat. <www.engineeringtoolbox.com>.

The Jakarta Globe, 2015. Indonesia to cut subsidised fuel prices. <www.thejakartaglobe.beritasatu.com>.

Wikipedia, 2014. Latent heat. <http://www.en.wikipedia.org/wiki/Latent_heat>.

CHAPTER 17

Oven Inspection and Audit

Contents

It is recommended that each oven is given a complete inspection and audit every 12 months. The annual inspection and audit will support the preventative maintenance program and help to ensure excellent oven performance, economy and long life.

17.1 OVEN PERFORMANCE

17.1.1 Output

- Check product specification and baking time to calculate oven output capacity in kg/hour
- Check actual output of packed, saleable product over 8-hour shift (Figure 17.1)

Figure 17.1 Packing table with manual biscuit transfer to packaging.

- Check downtime/causes/trends in oven output
- A log should be maintained of problems causing downtime. Any repetitive problems or trends in output can be identified and appropriate action taken.

17.1.2 Product Specification/Compliance

Check 10 biscuits from each lane every hour for compliance to the product specification.

Check variation across the width of the oven and variation with time:

- Biscuit weight
- Biscuit size (length, width, diameter)
- Colour (compare to standard samples)
- Moisture content

Identify any quality problems with Quality Control staff, such as 'checking', shelf life and packaging issues (Figures 17.2–17.4).

Figure 17.2 ESA Precision Scale from Brecknell Scales.

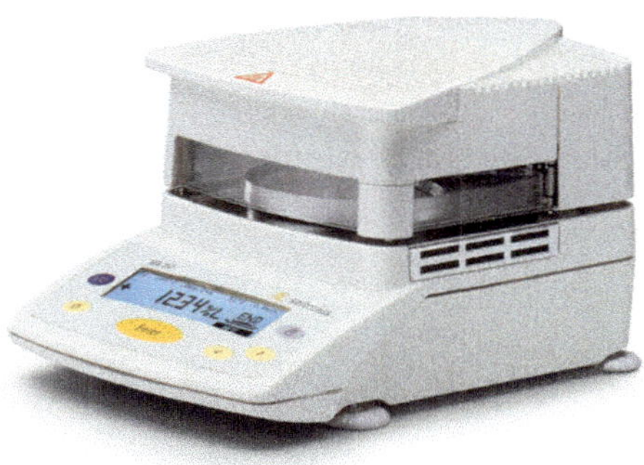

Figure 17.3 Sartorius moisture analyser MA.

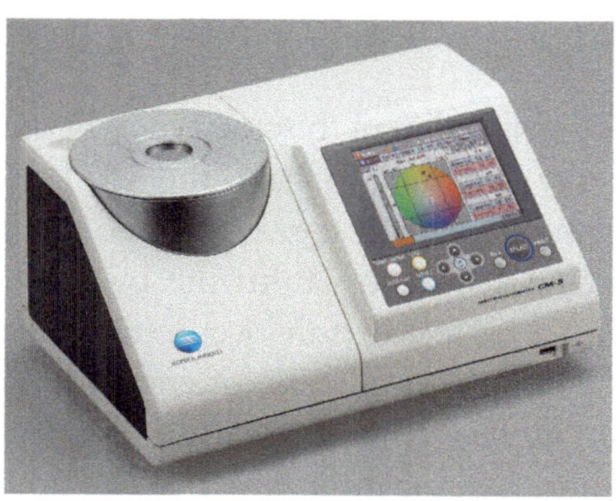

Figure 17.4 Konica Minolta CM-5 colour spectrophotometer.

17.1.3 Energy Usage

Check fuel consumption over the 8-hour shift (gas/oil usage).

Calculate energy in kW/kg of baked biscuits and compare with target consumption (Figures 17.5 and 17.6).

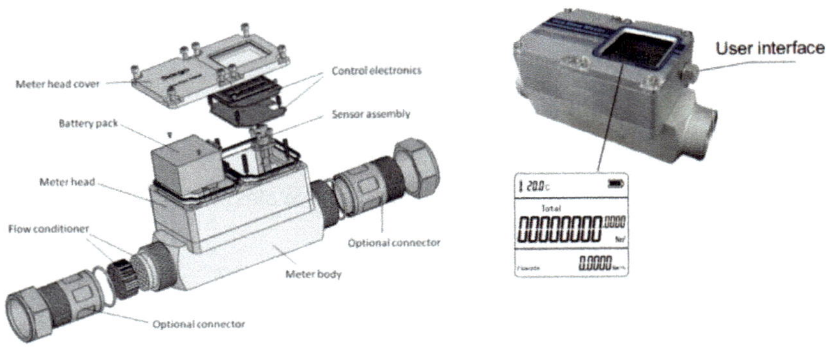

Figure 17.5 Gas meter: Bell Flow Systems.

Figure 17.6 Fuel meter: Great Plains Industries.

Energy usage for different biscuit types is given below as a guide. The 'energy required to bake the product' is the energy requirement for heat input to convert the dough pieces to baked biscuits; this excludes heat loss from the oven. The 'total energy required' assumes a typical biscuit oven efficiency including heat loss. The final column 'typical range of energy required' is a guide for comparison with actual energy usage. If the energy used exceeds this range, investigation is needed to improve the oven efficiency (Table 17.1).

Table 17.1 Energy usage guide

Biscuit type	Example	Energy required to bake product (kW/kg)	Total energy required includes heat loss from the oven	Typical range of energy required for baking (kW/kg)
Rotary moulded	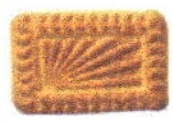	0.18	0.36	0.36–0.40
Hard sweet		0.25	0.50	0.50–0.55
Crackers		0.35	0.70	0.70–0.77

17.2 OVEN OPERATION AND MAINTENANCE

17.2.1 Oven Band

Check the following:

- Band condition (edge damage, cleanliness) (Figures 17.7 and 17.8)

Figure 17.7

Figure 17.8

- Band support skids/rollers (all rollers free and rotating, skids and rollers clean) (Figure 17.9)

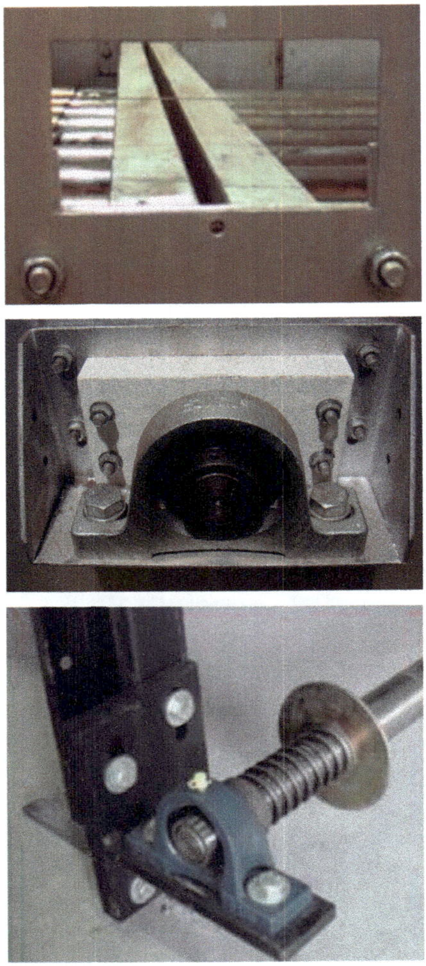

Figure 17.9 Oven band support skids and rollers.

- Band tracking (band wander limit switch settings to detect ± 10-mm band deviation)
- Check alarms and shut down at the set point (over 25 mm deviation) (Figure 17.10)

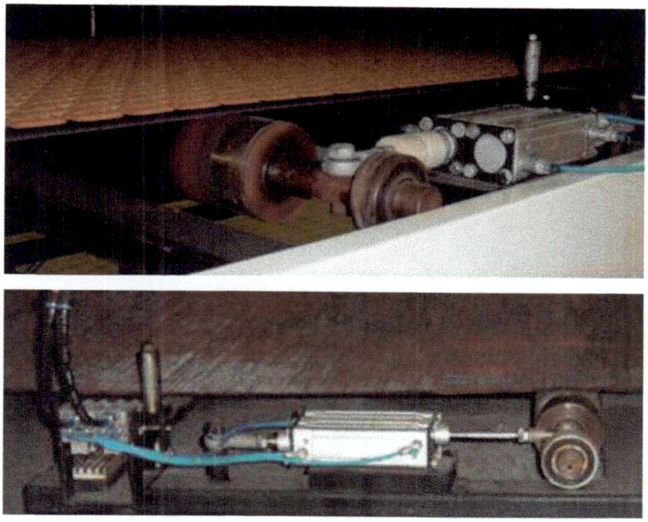

Figure 17.10 Band tracking systems.

- Band tension: check pneumatic system, air pressures, check actual tension applied compared to calculated requirement for the band type and oven length
- Check oven drum support bearings and slides. Check lubrication of slides (Figure 17.11)

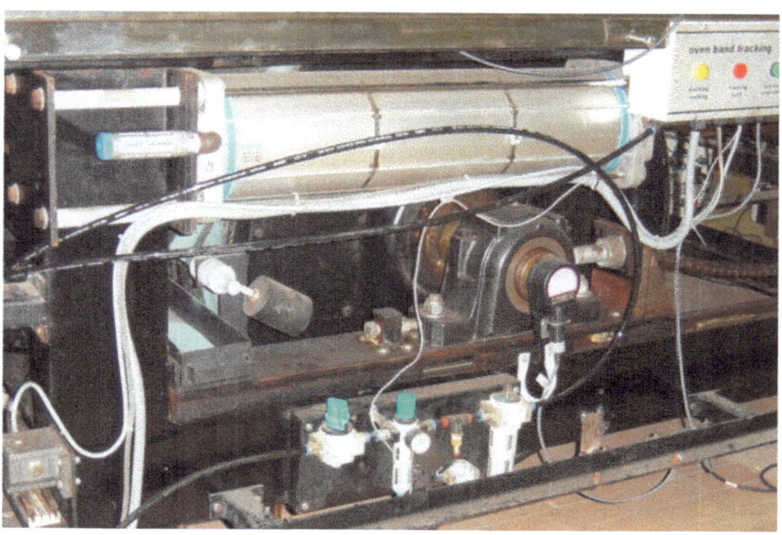

Figure 17.11 Oven tension end assembly.

- Oven drive: check belts and chains for wear and tension
- Check gearbox and drive transmission lubrication
- Check bearings (Figure 17.12)

Figure 17.12 Oven drive and transmission.

- Alarms: Check all alarm and safety circuits for band rotation, tracking and tension
- Emergency drive/uninterruptible power supply (UPS) (shut down power and check UPS system operates to empty the oven of biscuits) (Figure 17.13)

Figure 17.13 Oven emergency drive and batteries.

- Band cleaner (condition of brushes) (Figure 17.14).

Figure 17.14 Oven band cleaner.

17.2.2 Baking Chamber

- Length and zone configuration
- Insulation/temperatures of outer covers
- Check temperatures and heat loss at the oven ends (temperatures of outer covers should not exceed ambient + 10°C) (Figure 17.15)

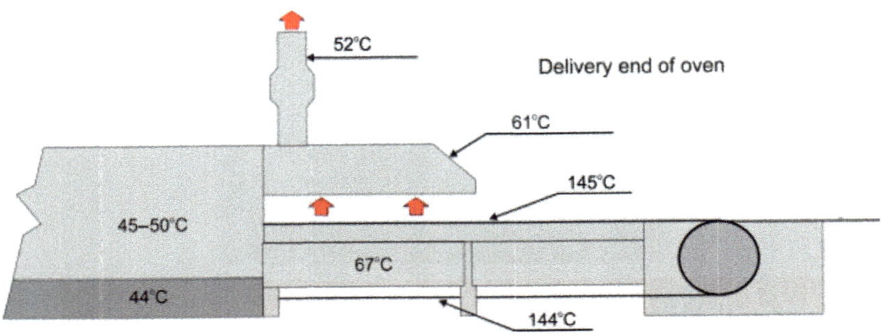

Figure 17.15 Example of diagram showing checks of the outer temperatures at the oven delivery end.

- Cleanout doors/condition inside the baking chamber
- Check for broken biscuits which could cause a fire
- Check for distortion and leaks in the baking chamber structure (Figures 17.16 and 17.17)

Figure 17.16 Cleanout door for indirect radiant oven.

Figure 17.17 Cleanout door for direct gas-fired oven.

- Extraction system/fans/pressure switches
- Check fan rotation, clean filters, bearings and impellers
- Check pressure switches setting and operation (Figures 17.18 and 17.19)

Figure 17.18 Extraction and combustion air fans.

Figure 17.19 Pressure switch.

- Oven end hood and extraction
- Check extraction fan/heat loss around delivery end (Figure 17.20).

Figure 17.20 Oven delivery end.

17.2.3 Gas/Oil Trains

See Figures 17.21 and 17.22.

Figure 17.21 Main gas supply train.

Figure 17.22 Weishaupt gas train for indirect radiant oven.

Check the following:

- Gas train installation/connections/gas filters
- Main gas supply pressure
- Main gas valve operation/pressure gauges/zero governor
- Gas proving system to detect any gas leaks
- Check the gas and air header pipes and flexible pipe connections (Figure 17.23).

Figure 17.23 Direct gas-fired oven: gas and air feeds to the burners.

17.2.4 Burners: Direct Gas-Fired Ovens

- Check burner strips for dirt/blockages
- Check ignition/flame monitor units
- Adjust electrode gaps
- Check flames: short blue flames (nor long, yellow, lazy flames)
- Adjust gas/air mix as required (Figures 17.24 and 17.25).

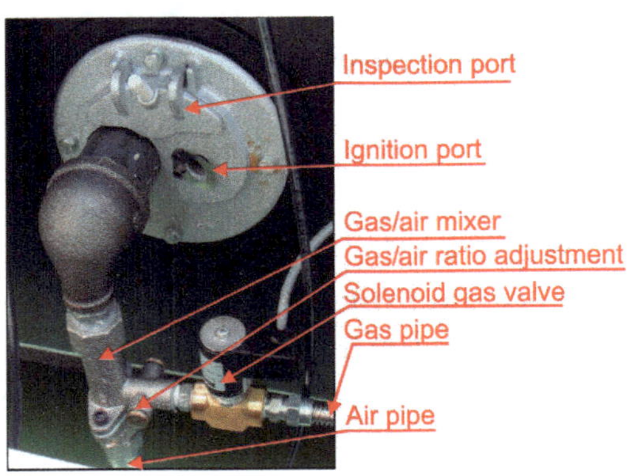

Figure 17.24 Flynn gas burner.

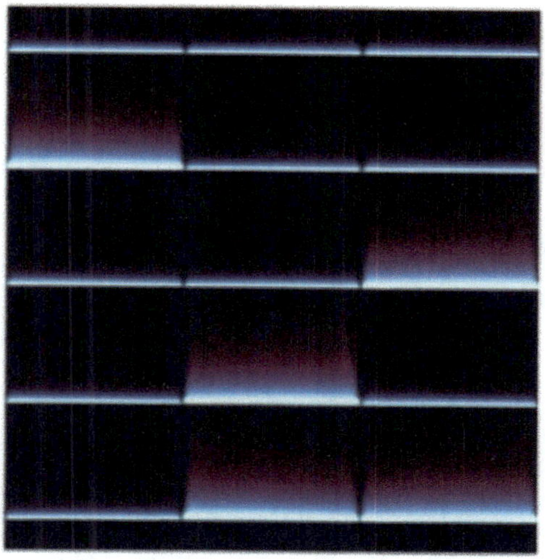

Figure 17.25 Flynn distributor burner adjustment.

17.2.5 Burners: Indirect Fired Ovens

See Figures 17.26 and 17.27.

Figure 17.26 Weishaupt gas burner.

Figure 17.27 Maxon dual fuel burner installation.

- Check fuel supply/filters
- Check combustion air fan/filter/pressure switches
- Check operation of manual and solenoid valves
- Check gas governor/gas pressure gauges
- Check pressure in burner tube with a manometer (should be negative pressure/positive pressure may indicate a leakage in the heat exchanger)
- Check burner ignition
- Check flame shape and stability
- Check Weishaupt Combustion Manager/diagnostics.

17.2.6 Temperature Control Systems

A data logger may be used to check the temperature profile of the oven and identify problems of variation across the width of the oven and variation from set temperatures in each zone.

17.2.7 Measuring Baking Profiles

In order to monitor actual baking temperatures, a thermal data logger is used. These units are slim and will pass through the oven carried on the oven band (Figure 17.28).

Figure 17.28 ThermaPro data logger from Digitron.

Data loggers are available with multichannels, usually six to eight for biscuit oven applications.

Each channel has a thermocouple sensor which is attached to a bar spanning the width of the oven band. In this way the baking temperature is monitored and recorded at multiple positions across the width of the oven.

Data loggers such as the TCK2000, which is made in China, monitor and record baking temperatures. A TCK system comprises the following: the data logger or profiler unit with three up to nine channels within an insulated box, thermocouples with high temperature connections, all accessories and software to run on a PC. This unit is a convenient way to monitor oven performance, detect problem areas in the oven which may cause poor structure, colour and moisture level (Figure 17.29).

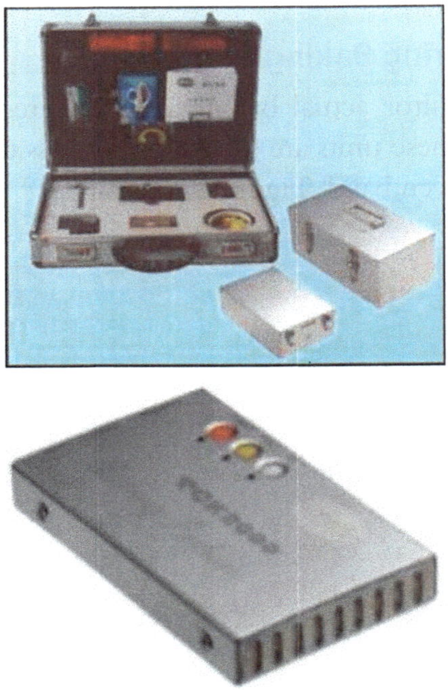

Figure 17.29 TCK 9 channel data logger to monitor and record baking temperatures.

Our main baking control in most ovens (direct gas-fired, indirect radiant and convection ovens) is by temperature. We set zone temperatures to match the required baking profile and control the heat input from the burners to maintain the set baking temperatures. Monitoring the baking temperatures through the oven and across the width of the oven band is therefore a valuable guide to the performance of the oven and a good starting point for troubleshooting problems such as uneven colour and moisture content.

17.2.8 Heat Flux

However, heat transfer depends not only on temperature, but is also affected by other conditions such as air movement. In an oven the heat energy will be transferred to the dough piece by radiation, conduction (from the oven band) and convection. In principle, heat flux sensors measure radiant, conducted and convective heat transfer. This gives a more complete understanding of the rate of heat transfer to the dough pieces than measuring temperature alone.

Heat flux is the rate of heat energy transferred to a given surface and it is measured in watts per square metre (W/m^2) ($1 W/m^2 = 0.86 kcal/hour/m^2$). This is expressed as:

$$Q'' = Q/A$$

where
Q″ = Heat flux (W/m^2)
Q = Heat transfer rate (W/hour)
A = Area (m^2)

17.2.9 Measurement of Heat Flux

Data loggers, such as the ThermaFlux from Digitron, record the rate of heat transfer, taking into account temperature, air velocity and conduction. They utilise the direct method of heat flux sensing (surface-mounted heat flux sensors).

When heat passes through the heat flux sensor, a temperature gradient is developed across it. The difference in temperature is detected by a thermopile in the heat flux sensor. The sensor head is connected to a data logger with a high memory capacity and can measure up to four samples/second (Figure 17.30).

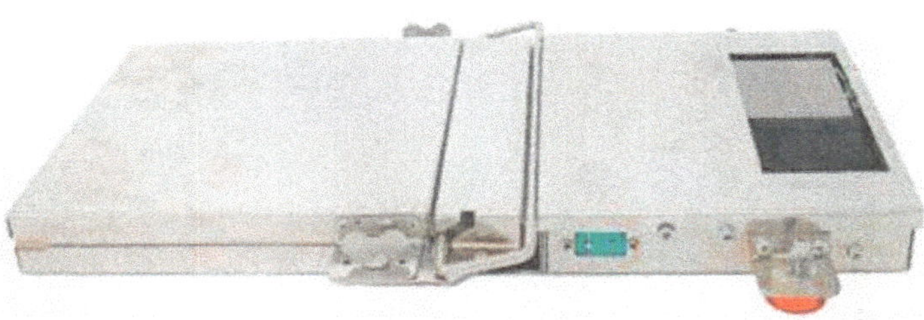

Figure 17.30 ThermaPro data logger for measurement of heat flux.

The data logger is slim, 25 mm, lightweight and portable with battery power. It has a dual sensor head enabling top and bottom heat flux measurements to be made simultaneously with recording air temperature.

The data logger will produce a graph showing both air (baking) temperature and heat flux at top and bottom of the sensor. Normally, it can be seen that the air temperature shows a (relatively) continuous curve. However, in a convection oven, the heat flux traces are more complex. They show very low dips at the end of each zone and a spiky trace reflecting the convection air jets impinging on the sensor through each oven zone.

17.2.10 Controls and Electrical Panels

- Check all temperature controllers are operational, including over-temperature controllers

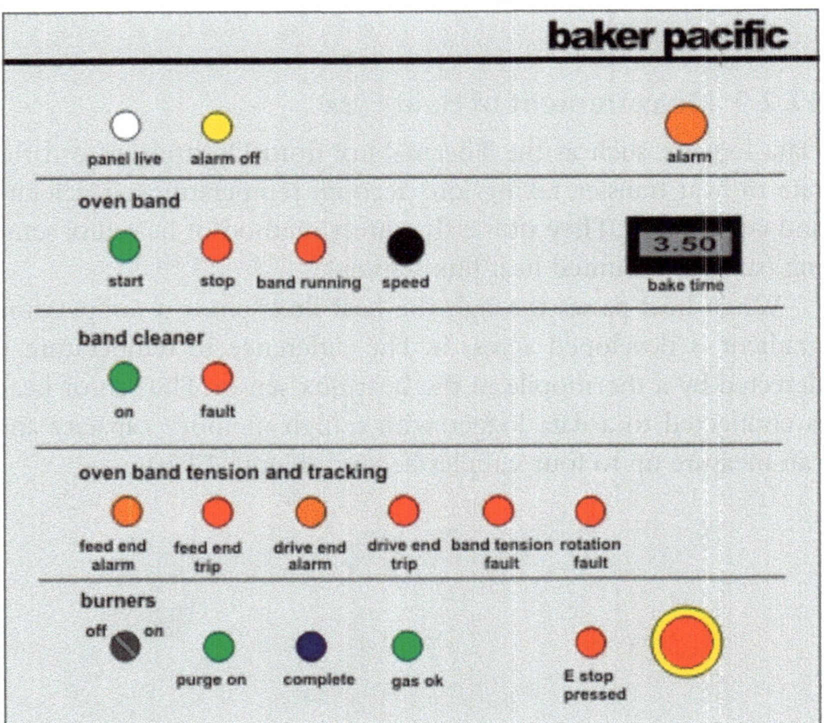

Figure 17.31 Main control panel.

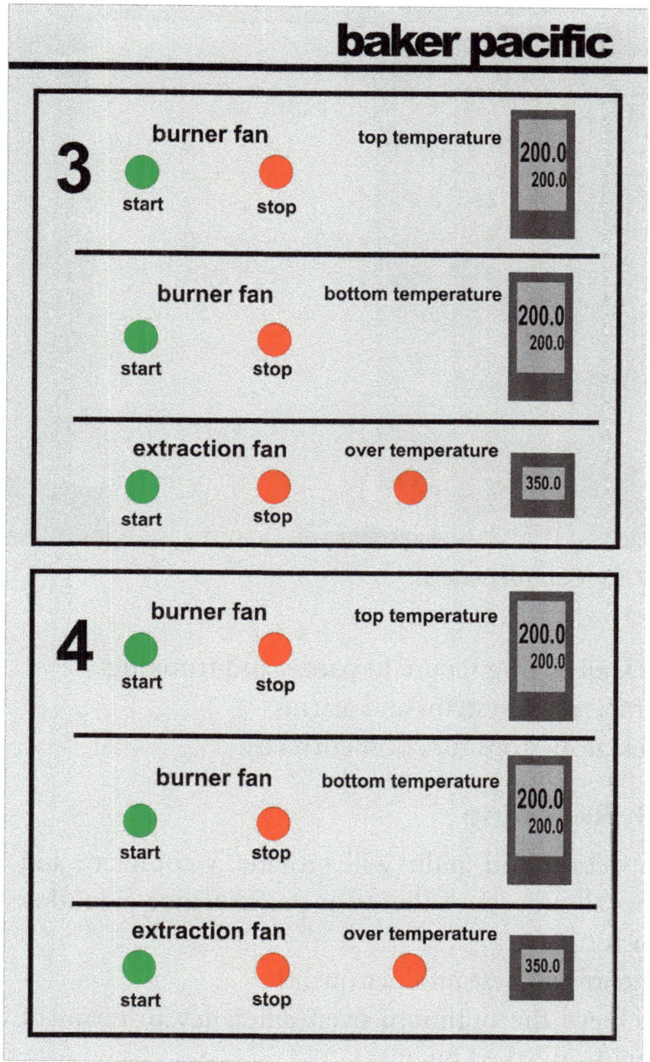

Figure 17.32 Zone control panel.

- Check variation in set temperatures and actual temperatures on the controllers
- Check proportional-integral-derivative settings and adjust as required
- Check thermocouple positions and wiring
- Check all damper controls (Figures 17.31–17.33)
- Check cleanliness and vacuum

Figure 17.33 Control panels.

- Check all wiring secure in panels and trunking
- Operate safety systems and alarms
- Check all E-stops function correctly.

17.2.11 Reporting

The inspection and audit will provide a complete and detailed report on all aspects of the oven performance as outlined above. The aim is:

1. To secure the best product quality
2. To achieve the optimum oven efficiency in terms of cost per kilogram of baked biscuits
3. To reduce downtime and achieve the highest possible production efficiency
4. To support development of preventative maintenance and a program of equipment and operational improvements.

BIBLIOGRAPHY

Baker Perkins Ltd, <www.bakerperkins.com>.

Digitron, 2015. Thermaflux, Thermal Profile Solutions, Data loggers for the direct measurement of Heat Flux. <www.digitron.co.uk>.

Eratec, 2015. <www.era-tec.fr>.

Flynn Burner Corp. USA, 2015. <www.flynnburner.com>.

Haas-Meincke, 2015. <www.haas.com>.

Maxon Corp, 2015. <www.maxoncorp.com>.

Weishaupt Corp, 2015. <www.weishaupt-corp.com>.

Wikipedia, 2015. Heat flux. <http://en.wikipedia.org>.

APPENDIX 1

Ingredients for Biscuits: An Introduction

A.1 WHEAT FLOUR

The flour shall be milled from good quality soft wheat, free from infestation and impurities, especially pesticides. The flour shall not be treated with enzymes and shall not contain chemical additives. The appearance shall be a good white colour, free from bran particles.

The odour shall be free from mustiness and foreign odours.

Two flours are typically used, described in the recipes as strong and weak. An analysis of each is given below.

	Weak/medium flour Semisweet doughs	Strong flour Fermented doughs
Starch	74.5	71.5
Moisture	14.0	13.5
Proteins (gluten forming)	7.0	10.0
Proteins (soluble)	1.0	1.0
Sugar	2.0	2.5
Fat	1.0	1.0
Ash (mineral salts)	0.5	0.5
Total	**100.0**	**100.0**

A.2 CORN FLOUR

Corn flour is a white free flowing powder produced by wet milling of maize, followed by washing, concentrating, centrifuging, drying, milling and sifting to give a natural maize starch. It has a short gel texture, relatively high viscosity and is easily dispersed in cold water.

A.3 SUGAR

The sugar shall be free from impurities and infestation. The appearance shall be a fine, white crystalline solid, free flowing and free from lumps.

The sugar shall have a sweet taste and be free from odours.

Moisture content	0.06% maximum
Ash content	0.03% maximum

Particle size: (1)

Powdered sugar	60 μm
Crystal sugar	150 μm
Caster sugar	150–450 μm
Granulated sugar	450–600 μm

Brown sugar, a dry golden brown sugar with bold crystals. The sugar should be dry and free flowing.

Particle size: 0.8–1.2 mm.

A.4 GLUCOSE SYRUP

Glucose syrup ($C_6H_{12}O_6$) is a solution (up to 80%) of glucose (dextrose), maltose and malto-dextrins in water. It is normally obtained by enzymatic hydrolysis of starch. Starch from wheat, corn, potato, cassava or any other plant can be used for this purpose.

Standard glucose syrup has a DE value (dextrose equivalent) of 42. The relative sweetness of 42DE glucose to sucrose is 40–45%.

A.5 CANE SYRUP 80%

Syrups with 80% solids derived from the refining of cane sugar. Used for their excellent flavour.

A.6 INVERT SYRUP 70%

Syrup of 70% solids made by acid hydrolysis of sucrose. The result is a 50:50 mixture of dextrose and fructose which are both reducing sugars and contribute to the Maillard reaction in baking.

A.7 FRUCTOSE SYRUP 80%

Commercially, fructose is usually derived from sugar cane, sugar beets and corn. Crystalline fructose is a monosaccharide, dried and ground, and of high purity.

High-fructose corn syrup is a mixture of glucose and fructose.

High-fructose syrup is a special type of glucose, formulated with a high fructose content. It is much sweeter than glucose.

A.8 MALT EXTRACT 80%

A thick glutinous syrup of 80% solids usually nondiastatic and obtained by water extraction of malted wheat or barley. The heat treatment used to concentrate the solution destroys any enzymes. It is used as an important flavour ingredient and rich in maltose, which is a reducing sugar.

A.9 DOUGH FAT

The fat shall be free from impurities and appear clean and bright when melted. The fat should be odourless, free from rancid and foreign flavours. The fat should be produced from good quality crude oils by a process of refining, bleaching and deodorising. It should be made primarily from vegetable oils, but it may contain hydrogenated fish oils.

Free fatty acid (as oleic acid)	maximum 0.08%
Peroxide value (milli equivalent per kg)	maximum 1.0
Moisture	maximum 0.1%
Slip melting point (BS 684 1.3 1976)	34°C–37°C

A.10 BUTTER

Butter is used for its shortening and flavour. The flavour of the butter is complemented by sugar and vanilla during baking and changes to a mild toffee or butterscotch flavour with good aroma.

Anhydrous butter or butter oil has almost no moisture and is harder than butter.

A.11 PALM OIL

Bleached, filtered and de-odorised.

Free from rancid or foreign smell.

Melting point: 36°C–38°C.

A.12 COCONUT OIL

For use in oil spray.

Hydrogenated coconut oil, neutralised, deodorised and bleached.

Melting point:32°C–34°C.

A.13 WHOLE EGG POWDER

Whole egg powder is spray dried. Egg is an ideal medium for the growth of micro-organisms, so great care must be taken to store and handle it with sterilised implements.

Egg yolk is rich in fat and lecithin and it is these ingredients together with the flavour, which contribute to the cookie.

A.14 LECITHIN

A complex natural surfactant obtained from soya beans.

A.15 YEAST (FRESH)

A microscopis, unicellular organism. It breaks down sucrose and maltose into monosaccharides and glucose and fructose into alcohol and carbon dioxide. During fermentation, the yeast will cause gluten modification and flavour development.

A.16 AMMONIUM BICARBONATE ("VOL") $(NH_4)HCO_3$

A volatile salt, which, when heated liberates carbon dioxide, ammonia gas and water.

A.17 SODIUM BICARBONATE

Sodium bicarbonate is the most important aerating agent. When heated it liberates carbon dioxide and water, leaving sodium carbonate as the residual salt. Sodium carbonate has a softening action on gluten and darkens the biscuit.

If sodium bicarbonate is heated, only half the carbon dioxide is released, but if an acid is present all the carbon dioxide is released and there is no softening action on the gluten or darkening of the colour of the biscuit.

A.18 ACP – ACID CALCIUM PHOSPHATE

Acid calcium phosphate is also known as monocalcium phosphate. It is fairly soluble in cold water, but for doughs which are used without standing, a good proportion of the reaction takes place during baking. It is used in conjunction with sodium bicarbonate and ammonium carbonate.

A.19 SAPP – SODIUM ACID PYROPHOSPHATE

Sodium acid pyrophosphate is commonly used in the baking industry as a leavening agent. It combines with sodium bicarbonate to release carbon dioxide.

A.20 SALT

Sodium chloride is used as a flavour enhancer and also to control the rate and extent of fermentation.

A.21 SMS – SODIUM METABISULPHITE $NA_2S_2O_5$

A reducing agent for the modification of the strength of the gluten in doughs.

A.22 PROTEOLYTIC ENZYME

Used to soften the gluten in doughs. A white, fine powder.

Recommended source: Kerry Bioscience.

BIBLIOGRAPHY

Baking Management, 2008. A tropical touch. Soy-based solutions. Penton Media Inc. <http://www.baking-anagement.com/ingredients/brief_guide_functional_1028/index1.html>.

Benedict, M. University of Houston. How does temperature affect yeast activity? MadSci Network. <http://www.madsci.org/posts/archives/jan2001/980908832.Gb.r.html>.

British Sugar plc, 2015. Screened White Sugars. <www.britishsugar.co.uk>.

Buck, JS, Walker, CE, Sugar and sucrose ester effects on maize and wheat starch gelatinisation patterns. Starch. Wiley VCH Verlag GmbH. 2009.

Dakota Yeast, 2010. Yeast Fermentation in Baked Goods. <http://www.dakotayeast.com/help-fermentation.html>.

DTKFCPL. Examining the Role of Fats in Bakery. APR03.

Eyre, C., 2008. AB Enzymes Launch Targets Improved Biscuit Baking. Bakeryandsnacks.com Decision News Media. <www.bakeryandsnacks.com/content/view/print/131574>.

Fineli®. National Institute for Health and Welfare. Wheat Flour Whole Grain. 2003–2010.

Flour Specifications, 2015. <www.flourmilling.co.uk>.

Food Resource, 1990. Oregon State University. Gel. Excerpts from Bender Arnold E. Dictionary of Nutrition and Food Technology, Butterworths, Boston.

Food Resource, 2010. Oregon State University. Bread Dough. Starch.

Ghiasi, K., Hoseney, R.C., Varriano-Marston, E., 1981. Effects of flour components and dough ingredients on starch gelatinisation. Cereal Chem. 60 (1), 58–61.

Gurney, A. Bakery Fats and Oils. Leading Edge. 2008.

Lansbergen, G., 2002. Fats for Food Consultants. <http://www.fatsforfoods.com/specifications.htm>.

Lowe, B. Experimental Cookery from the Chemical and Physical Standpoint. Gluten. 2009.

Manley, D., 2001. Biscuit, Cracker and Cookie Recipes for the Food Industry. Woodhead Publishing Ltd.,Cambridge CB16AH, UK, ISBN 1 85573 543 1. CRC Press LLC, FL33431, ISBN 0 8493 1220 5.

Moodie, P. Traditional Baking Enzymes – Proteases. Enzyme Development Corporation. Presented at American Institute of Baking, Manhattan, Kansas. 2001.

Toufeili, I., Ismail, B., Shadarevian, S., Baalbaki, R., Khatar, B.S., Bell, A.E., Schofield, J.D., 1999. The role of gluten proteins in the baking of arabic bread. J. Cereal Sci. 30, 255–265.

APPENDIX 2

Oven Maintenance

- Before carrying out any maintenance, ensure that the equipment is isolated and that all power and fuel supplies are shut down:
  - *Electrical supply*
  - *Pneumatic supply*
  - *Fuel supply*
- Ensure that all lock out devices have been activated
- Ensure that the oven has cooled down to a safe working level.

A2.1 NEW EQUIPMENT

After the oven has been running for one month, check that all the nuts, screws, setscrews, taper pins, etc are tight. These should then be checked over periodically. Include all electrical terminals in these checks (Figure A2.1).

Figure A2.1 Baker Pacific direct gas fired / indirect radiant oven.

A2.2 SAFETY DEVICES

All safety devices must be checked for correct operation at weekly intervals.

A2.3 ROUTINE MAINTENANCE

Regular maintenance is essential to prevent expensive stoppages of production. Baker Pacific recommends that regular preventative maintenance be carried out and any items requiring attention dealt with as soon as possible to prevent any further damage. Maintenance schedules are set out below.

A2.4 MECHANICAL COMPONENTS

A2.4.1 Welded Components

Where components are subject to stress and load, it is important that they are checked regularly (weekly) for integrity and general condition of the welds.

A2.4.2 Driving Chains and Belts

Should be checked regularly (weekly) for general condition and tension, adjustment being made if necessary.

A2.4.3 Motors and Drives

Regular attention (monthly) should be paid to the mountings to ensure they have not worked loose, and that the drive alignment is correct. Adjust and tighten if necessary.

A2.4.4 Steam Lines and Fittings

Particular care must be taken to ensure any steam leaks are located and repaired as soon as possible. Steam is dangerous and may injure personnel working in the area.

A2.4.5 Air Lines and Fittings

Regular inspection (monthly) of the pneumatic system should be made and any leakages corrected. Any air lubricators should be topped up with lubricant and filters should be checked and cleaned.

A2.4.6 Seal and Gaskets

Replace any seals or gaskets that show signs of deterioration or damage. When assembling, ensure that any nuts etc are tightened evenly so that the seal and housing are drawn square to the shaft.

A2.4.7 Bearings

When replacing a bearing ensure that it is correctly lubricated. Note that many bearings are sealed units requiring no further lubrication.

A2.4.8 Conveyor Belts

Should be checked regularly (weekly) for general condition. The belt tension and belt tracking should also be checked at the same time to ensure maximum belt life. Where automatic belt tensioning and tracking systems are fitted they should be checked for correct operation.

A2.4.9 Oven Band Tension

The oven band tension and tensioning system should be checked every week.

A wire-mesh band should be only just taut enough for the drum to drive it; excessive tension will stretch the band unduly and cause a reduction in width.

Check each week while the oven is at baking temperature, that the limit of travel of the feed end drum has not been reached and that no slip occurs on the driving drum at delivery end.

Eventually, a wire band will stretch up to its final operating length. Before the drum reaches the limit of its travel the band must be shortened by cutting out a section and remaking the joint.

A2.4.10 Pneumatic Tension Arrangement

The movement of the feed end drum is controlled by air cylinders. A pre-set air pressure acts on the cylinder pistons and keeps the band at the correct tension at all times.

Incorporated in the pneumatic systems is a relief valve which will relieve the excessive pressure that tends to build up at the end of a production run as the oven cools and the band contracts. A nonreturn valve is also fitted to prevent the fall of cylinder air pressure should the main air supply fail.

A2.4.11 Band Pressure Switch

An air pressure switch ensures that air at the correct pressure is available before the oven drive can be started.

The air pressure regulator, relief valves and pressure switch have been set by the commissioning engineer to suit the size of oven and type of the band. No subsequent readjustments should be necessary.

A2.4.12 Oven Band Tracking

Automatic tracking rollers are fitted at each end of the oven before the end drums. These are pneumatically operated when the band wanders and engages one of the limit switches. The band wander warning switches and the free movement of the assembly should be checked weekly. The pneumatic equipment and free movement of the tracking rollers should also be checked weekly.

The vertical side guide rollers have been set clear of the band and there should never be continuous contact between the band and the guide rollers – otherwise the edge of the band may be damaged.

After the initial running and commissioning period, very small adjustments to two or three rollers supporting the return band under the oven should be sufficient to track the band correctly. Any tracking adjustments should be made with the band running at slow speed and it will take two or three complete revolutions of the band before the adjustments take effect.

A2.4.13 Bearings for Oven Band Support Rollers

Each oven band roller is supported on pillow block bearings. Check the bearings regularly for free running.

A2.4.14 Temperature Controllers

It is recommended that arrangements are made for the periodic servicing of the equipment.

A2.4.15 Motors and Electrical Control Equipment

These items must be cleaned and serviced and adjusted at monthly intervals by a competent electrician as detailed in the manufacturer's instructions.

A2.4.16 Pulleys for Fans

Fan and motor pulleys which are fitted with taper-lock securing bushes should be checked with reference to the supplier's instruction sheet for details of removal and fitting instructions.

A2.4.17 Fans

Any tendency to noisy running should be investigated. It may be due to bearings, which should be replaced or lubricated, or to dirt build-up on the impeller, which should be cleaned.

A2.4.18 Explosion Door Arrangement

Explosion doors are incorporated in the oven to minimize the damage of an explosion.

Every year check the explosion panels and replace any damaged parts.

A2.4.19 Oven Flues

Periodically inspect the exterior sections, roof fixings, weather capes and rain hoods for security and state of repair.

Periodically inspect the flanged joints to ensure that a gas tight seal is being maintained.

Renew any defective joints, using a standard joint ring made from a compressed fibre.

Should welding be necessary due to repairs or modifications to the equipment, use an anti-acid grade stainless steel type of rod.

A2.4.20 Battery Safety

(See manufacturers' data)

The gases that can be emitted are explosive and the battery area must therefore be well ventilated. All forms of external ignition, such as gas torches and welding equipment, must not be used in the vicinity of batteries. Always wear safety goggles when working on batteries.

Battery acid on the skin must be washed with liberal amounts of clean water. A dilute sodium bicarbonate solution can be used as an eye-wash for sulphuric acid. For potassium hydroxide, use a weak solution of vinegar. Otherwise, bathe with fresh water. Always consult a doctor after using an eye-wash.

Cells and batteries are always live, even when the electrolyte has been totally lost through container breakage or spillage. Keep metal objects well away from cells as sparks may cause an explosion.

Insulated spanners must be used when working on a battery. Do not remove a cell from the circuit without first isolating the load and charger.

A2.5 ELECTRICAL MAINTENANCE

A2.5.1 General

Components with moving parts should be inspected at intervals related to their frequency of operation.

Pitting of electrical contacts after a short period of service is normal bedding-in and is not detrimental to their operation. Modern contacts have a special finish and must not be 'dressed'. Silvered contacts will oxidize and become black and should not be polished since this would remove some of the coating.

A2.5.2 Cleaning

Every six months, control panels should be cleaned using a vacuum system. If this is not available, use a hand blower or a compressed air line. Avoid excessive use of compressed air because entrained moisture could be harmful.

A2.5.3 Relays

The normal mechanical life of relays and contractors is approximately five million operations. Baker Pacific therefore recommends that those components which are subjected to frequent switching be checked and, if necessary, replaced after approximately five million operations.

Relays not subjected to frequent operation, for example, those in alarm and safety circuits should be inspected and tested regularly (weekly).

A2.5.4 Inspection Schedule for Relays

Table A2.1

A2.5.5 Other Electrical Equipment

A2.5.5.1 Connections and Leads

Every six months, check all terminals for tightness, security and that they are safe. Check that all insulating and safety shrouds are securely in place and in good condition.

Table A2.1 Inspection schedule for relays

Type	Frequency	Action
Low duty	Operated less than once an hour	Inspect every 6 months
Normal duty	Operated more than once an hour but less than once per minute	Inspect every 3 months
High duty	Operated more than once a minute	Inspect every month
Safety duty	N/A	Inspect and test weekly

Check that flexible cables have freedom of movement, particularly on devices which move frequently.

Check that the insulation of cables has not been cracked or worn by moving parts, either mechanical or by opening doors or covers.

A2.5.6 Fuses

If one power fuse breaks on a three phase supply then the other two should be renewed also.

Always replace fuses on a three phase systems as a set of three.

A2.5.7 Earth Leakage Protection Devices

Where these are installed a regular (monthly) operation check should be carried out.

A2.5.8 Clutches and Brakes

Refer to the manufacturer's instructions for inspection requirements with due regard to the frequency of operation.

A2.5.9 Transistor/Solid State Devices

The life of transistor and solid state devices is indefinite. In the event of failure of a transistor device, circuit or component, replace the complete unit or plug-in board.

A2.5.10 Limit Switches

The actuator should be regularly inspected (monthly) for freedom from wear or damage. The operation and alignment should be checked and adjusted if needed. Inspect the rod or roller and any rubber seal around the plunger actuator. Renew the switch after approximately one million operations.

A2.5.11 Proximity Detectors (Inductive or Capacitive)

These are solid state devices and require no periodic maintenance, except to check that the mounting is secure with the correct sensing gap and to wipe them clean. The frequency of cleaning will depend upon operational conditions.

A2.5.12 Plug-In Timers

The plug-in timer is a transistor unit and contains a relay similar to a normal plug-in relay with similar life and maintenance requirements.

A2.6 MAINTENANCE SCHEDULE

A2.6.1 Each Day

Drain the filter bowl of the pneumatic band tensioning arrangement.

Empty and clean the scrap trays at the feed and delivery ends.

Check the oven band tracking and tension equipment.

A2.6.2 Each Week

A2.6.2.1 Driving Belts

Carefully examine all driving belts for correct tension and wear. Refer to maker's instructions for details.

A2.6.2.2 Fan Filters

Remove and wash the combustion air fan filters. Replace when dry.

A2.6.2.3 Inspection and Access Doors

Examine the seal between the door and its frame and rectify any defect.

A2.6.2.4 Oven Lamps

Keep the oven lamp glasses clean and regularly examine the bulb and replace it if it becomes discoloured. Examine the holder and the flexible cable.

A2.6.2.5 Dampers

Operate every heat control and extraction damper throughout its full working range to ensure that they work freely.

A2.6.2.6 Emergency Drive

Run the emergency drive and check the UPS system.

A2.6.2.7 Oven Ignition

Inspect and reset the spark gap to 3.5 mm. Check for cracks in the ceramics.

A2.6.3 Every Twelve Months

A2.6.3.1 Burners and Gas Equipment

Zero Pressure Regulator – Inspect diaphragms and valve. Check that the impulse tube is not blocked.

- Have the burners serviced by a specialist.
- Remove any build-up of dirt within the burner nozzles.
- Check flexible air and gas hoses for any leakage. Replace where necessary.
- Inspect high tension ignition leads. Replace if necessary.

A2.7 OVEN CLEANING

A2.7.1 Inside the Oven

CAUTION: DO NOT ATTEMPT TO CLEAN INSIDE THE OVEN

Regular cleaning of the inside of the oven must be carried out as a build-up of material can constitute a fire or explosion hazard.

The oven is designed for easy cleaning using an industrial vacuum cleaner with a flame-proof motor.

Check oven band vibration, especially at the delivery end.

Toppings applied before baking, such as salt, sugar etc. should be carefully monitored and recovery systems checked.

A2.7.2 Under the Oven

Regular cleaning with an industrial vacuum cleaner is required.

A2.7.3 Outer Surfaces of the Oven

The top should be vacuumed periodically. All outer sheets should be wiped down periodically. Use a damp cloth and do not wash as water may seep into the insulation.

A2.7.4 Motor and Fan Covers

These should be periodically vacuumed. Do not wash as water may seep into the motor or fan components.

A2.7.5 Gas Filters

A filter may be incorporated in the gas supply to each burner and in the supply to the oven. Check regularly and clean or replace.

Cleaning of the element can, in most cases, be carried out satisfactorily by the use of compressed air.

A paper filter will require replacement at periodic intervals. If the element is of metal gauze, it may be washed in paraffin. Care should be taken not to damage the felt seals at either end of the element that otherwise could require replacement.

A2.8 STANDARD LUBRICATION

A2.8.1 Ball/Roller Bearings With Provision for Lubrication

Packed with grease at assembly, these bearings need little further attention other that the occasional addition of a small amount of grease. If examination becomes necessary, clean and flush out the bearing with an environmentally safe, approved quick drying solvent and if the bearing is serviceable, repack with new grease.

A2.8.2 Sealed Ball Bearings

This type of bearing is lubricated for the life at the time of manufacture and no provision is made for lubrication. If the bearing is running noisily it should be replaced.

A2.8.3 NSK – RHP Self Lube Bearings

These bearings are factory charged with the correct type and quantity of grease at the time of manufacture and seldom require recharging.

If the bearings do however require more grease (eg, due to excessive loads or operating temperatures), they can be recharged with two or three shots of grease chosen from the Recommended Lubricants chart.

A2.8.4 Recommended Lubricants

Reference should be made to the supplier's lubrication information.

A2.9 GENERAL LUBRICATION SCHEDULE

The required frequency of lubrication checks and changes is dependent upon the operating conditions of the equipment. The following schedule is given as a guide, which will be appropriate for most installations (Table A2.2).

Table A2.2 General lubrication schedule

Item	Frequency	Lubricant
Heavy duty gears	Every day	Lubricate/grease
Driving chains	Every week	Oil as required
Light duty gears	Every week	Lubricate/grease
All gearboxes	Every week	Check oil levels and replenish if necessary
Fan and motor bearings	Every three months	One or two shots only of the appropriate grease
All gearboxes	Every six months	Change the oil
Bronze oil-retaining bushes	Every six months	Apply oil liberally, allow some to be absorbed by the bush, then clean off the surplus oil

A2.10 MAINTENANCE LOG OR RECORD

It is essential for the user to record all maintenance activities on the equipment. Such a log will assist in identifying any persistent problems, keep track of consumable and spares items used and also act as a means of communication from one group of maintenance engineers to another. A sample chart for this log is shown below (Table A2.3).

Table A2.3 Maintenance log: Line No: ....

Details of work done	Date	Signature

A2.11 RECOMMENDED SPARE PARTS

It is recommended that the factory has a stock of key spare parts. Failure of key parts can cause one or more zones of the oven to be shut down. The following parts should be readily available to the maintenance engineers.

A2.11.1 Gas System

- Main gas supply manual shutoff valve
- Main gas supply solenoid valve
- High and low gas pressure switches for the gas train
- Spare Dungs zero gas governor for direct gas-fired oven.

A2.11.2 Weishaupt Burners

- Dungs high and low pressures switches
- Gas shutoff valve for Weishaupt burner.

A2.11.3 Maxon Burners

- Landis & Gyr Maxon burner control unit
- Flame detector/flame eye for Maxon burner
- Combustion fan air pressure switch for Maxon burner
- Ignition rod for Maxon burner.

A2.11.4 Flynn Burners

- Flynn ignition control units
- Flynn spark ignition electrodes
- Gas solenoid valves for Flynn burners.

A2.11.5 Electrical/Temperature Control Parts

- Six-minute fixed purge timer
- Antunes pressure switches for the circulating and extraction fans
- Bake time/band speed proximity sensor (Synatel)
- Selection of thermocouples to suit all ovens
- Thermocouple high temperature compensating cable

- Omron oven temperature controller
- Omron over temperature display unit
- A range of spare electrical motor inverters.

A2.11.6 Mechanical Parts

- Range of belts to suite all type of fans
- Complete range of bearings for all drives, fans, rollers, etc.
- Spare stripping blade at oven transfer to stripping conveyor
- Spare brushes (and flapper material) for band cleaners.

BIBLIOGRAPHY

Baker Pacific Ltd. <www.bakerpacific.com.hk>.
Baker Perkins Ltd. <www.bakerperkins.com>.

APPENDIX 3

Combustion Data

The combustion process is a reaction of rapid oxidisation started by the correct mixture of fuel, oxygen and an ignition source. The chemical reaction for natural gas is:

$$CH_4 + 3O_2 = \text{Heat} + 2H_2O + CO_2 + O_2$$

Where:
CH_4 = 1 m^3 of natural gas
$3O_2$ = 3 m^3 of oxygen
$2H_2O$ = 2 m^3 of water vapour
CO_2 = 1 m^3 of carbon dioxide
O_2 = 1 m^3 of excess oxygen.

Air is composed of 20.9% of oxygen, 78% of nitrogen and 1% of other gases. For most applications, every 1 m^3 of oxygen, approximately 15 m^3 of air is required to provide complete combustion of natural gas. To ensure complete combustion of the fuel, excess air is drawn in by the burners. The combustion efficiency will increase with increased excess air, until the heat loss in the excess air is larger than the heat provided by more efficient combustion.

When fuel and oxygen in the air are in perfectly balance - the combustion is said to be *stoichiometric*.

Typical excess air to achieve the best efficiency for combustion is:

5–10% for natural gas

5–20% for fuel oil.

Carbon dioxide is a product of the combustion and the content in the flue gas is an important indication of the combustion efficiency. An optimum content of carbon dioxide after combustion is approximately *10%* for natural gas and approximately 13% for light fuel oils.

BIBLIOGRAPHY

The Engineering ToolBox, 2015. <www.engineeringtoolbox.com>.

Testo Inc. Applications Guide Rev. 1.0, 2006. <www.testo-international.com>.

APPENDIX 4

Oven Manufacturers

Contents

China

China Light Industry Machine Association
Dingson Food Machinery Ltd
Dongguan Furong Food Machinery
Evergrowing Food Machinery Co.
Foshan Shunde Huaji Industrial Co.
Skywin Foodstuff Machinery Co.
Zhuhai Hong Fu Mechanical Manufacturing Co.

Europe

Aasted
Baker Perkins Ltd
DFE – Meincke – De Vuurslag
Fiamma Baking Systems
Haas – Meincke
Ima Forni International
Laser srl
Polin
Spooner Vicars
TMFCT – Bonnand Lornac
Werner & Pfleiderer

India

Bake-o-nomic
Besto Oven Industries
Esspee Engineers
New Era Machines
Ovenman
Prima Engineering

Japan

Misuzu BP Corp
Naigai Vicars

Korea

Dong Yang Food Machinery Co.

USA

Food Process Automation
Reading Bakery Systems

CHINA

China Light Industry Machinery Association - CLIMA

Address: RM 3A07, YinGu Plaza, No 9 West Road, North Fourth Ring Road, Beijing 100080, China

www.clima.net.cn

Dingson Food Machinery Ltd

13 Teng Yun Road, Tanzhou Town, Zhongshan City, Guangdong Province, China.

www.dsm-mc.com

Dingson International Ltd

Unit A, 16/F, Teda Building, 87 Wing Lok Street, Sheung Wan, Hong Kong

export@dsm-mc.com

Dongguang Furong Food Machinery

Nanshe Industrial Chashan Town, Dongguan City, Guangdong Province, China, Zip Code: 523391

www.dg-furong.cn www.ftt.cn

Evergrowing Food Machinery Co. Ltd

Rm 418, LinHai Building, ShiHua Rd, ZhuHai, China
http://www.biscuitline.cn http://www.foodsmachine.net

Foshan Shunde Huaji Industrial Co.

China Huaji Industrial Co. Ltd

Kunzhou Industrial Zone, Beijiao Town, Shunde, Foshan City, Guangdong Province, China
www.sinobake.net

Skywin Foodstuff Machinery Co. Ltd

Joint-venture Industrial Area, West Zone, Yangjiang City, Guangdong, China
www.skywinbake.com www.skywinbakery.com www.skywin-machinery.com

Zhuhai Hong Fu Mechanical Manufacturing Co. Ltd

China Meihua Road, Xiangzhou Technology Industrial Park, Zhuhai City, Guangdong Province, China, 519071
www.hfmachine.cn

EUROPE

Aasted

Bygmarken 7–17, DK 3520, Farum, Denmark
www.aasted.eu

Baker Perkins Ltd

Baker Perkins Ltd

Manor Drive, Paston Parkway, Peterborough, PE4 7AP, United Kingdom
www.bakerperkins.com

Baker Perkins, Inc

3223 Kraft Ave. S.E. Grand Rapids, MI, 49512-2027, USA
www.bakerperkins.com

DFE Meincke – Vuurslag B.V.

Schelling Straat 2, 4879 NK, Etten-Leur, Nederland
www.vuurslag.nl

Fiamma Baking Systems S.r.l.

Via Meucci 8, 37045 Legnago, Italy
www.2fsrl.net

HAAS-MEINCKE A/S

Tonsbakken 10, 2740 Skovlunde, Denmark
www.haas.com

IMAFORNI INT'L S.p.A.

37030 Colognola ai Colli - via Strà, 158 - Z.I.Verona, Italy
www.imaforni.com

Laser Srl

Via Saturno 36, 37059 Santa Maria di Zevio –Verona, Italy
www.laserbiscuit.it

Polin

Ing POLIN & C. SpA
Viale dell'Industria 9, 37135 Verona – Italy
www.polin.it

Spooner Vicars Bakery Systems

Middleby House, Unit 15, Bridge Bank Close, Off Yew Tree Lane, Stonecross Park, Golborne, Wigan, WA3 3JD, United Kingdom
www.spoonervicarsbakery.com

TMFCT – Bonnand Lornac

Zone Industrielle, 26 rue de St Rémy, 28270 Brezolles, France
www.bonnandlornac.fr www.tmfct.fr

Werner & Pfleiderer

Industrielle Backtechnik GmbH, Frankfurter Straße 17, 71732 Tamm, Germany

www.wpib.de

INDIA

Bake-o-Nomic Corp

6 Kiran, K-R-N Society, Off S.V.P. Road, Borivli (West), Mumbai, Maharashtra 400092, India

1, M&n Estate, Opp Thakur Mall, Near Dahisar Check Post, Mumbai, Maharashtra 401104, India

wonderbake@vsnl.com

Besto Oven Industries

8, Acharya complex, Dr CG Road, Near Basant Cinema, Chembur, Mumbai, Maharashtra 400074, India

www.ovenindia.com

Esspee Engineers, Kolkata

137, Biplobi Rash Behari Bose Road, 2nd floor, Kolkata, West Bengal 700001, India

New Era Machines

New Era Machines Private Ltd.

Campa Cola lane, Dhandari Kalan, GT Road, Ludhiana 141010, India

Ovenman Industries Private Ltd

B 14-15 Industrial Area, Site 1, Haripur, Jalalabad, Lucknow Road, Faizabad 224001, India

www.ovenmanindia.com

Prima Engineering Industries

T -113, Near M.I.D.C. Post Office, Bhosari, Pune 411026, India
www.primaengg.com

JAPAN

Misuzu BP Corporation

Shiba Fuji Bldg 3F, 1-1-15 Shibadaimon, Minato-Ku, Tokyo, Japan

Misuzu Koki Co Ltd

2nd Fl Cosmic Building, 1-2-8 Shibadaimon, Minato-Ku, Tokyo, Japan
www.misuzukoki.co.jp

KOREA

Dong Yang Food Machinery Co. Ltd

301-10 Doksan 1 Dong, Gumchon-gu, Seoul, Korea 153-011
www.mixer.co.kr

USA

Food Process Automation

2450 Oak Industrial Drive, Grand Rapids, MI 49505, USA

Reading Bakery Systems

380 Old West Penn. Avenue, Robesonia, Pennsylvania, PA 19551, USA
www.readingbakery.com

APPENDIX 5

Oven Band Manufacturers

Contents

AGRATI GROUP LA BRIDOIRE S.A.R.L.

Produits Tréfilés de la Bridoire S.A.R.L.

A.Agrati S.p.A.

Via Piave 28-30, 20837 Veduggio con Colzano (MB) Italy
info@agrati.com; www.agrati.com

Agrati La Bridoire S.A.R.L.

640 Route du Lac, 73520 La Bridoire, France
info@agrati-france.fr

ARK ENGINEERS

Thakur Niwas 16-D, Samrath Mill Compound, LBS Marg, Vikhroli-W, Mumbai-400 079, Maharashtra, India

ASHTON GREEN & COMPANY

No. 118, Sir Vithaldas Chambers, 16 Mumbai Samachar Marg, Mumbai-400 001, Maharashtra, India
www.wiremeshconveyor.com

ASHWORTH BROS. INC.

450 Armour Dale, Winchester, VA 22601, USA
ashworth@ashworth.com; www.ashworth.com

AUDUBON

850 Pennsylvania Blvd., Feasterville, PA 19053, USA
www.meshbelt.com

BERNDORF BAND GMBH

Leobersdorfer Strasse 26, A-2560 Berndorf, Austria
www.berndorf-band.at

BHARAT WIRE MESH COMPANY

D-11, Shiv Bhole Laghu Udyog, Opp Hindustan Construction Company, Ambewadi, L.B.S. Marg, Vikroli (West), Mumbai-400 083, India
www.bharatwiremesh.com

CAMBRIDGE ENGINEERED SOLUTIONS

105 Goodwill Road, Cambridge, MD 21613, USA
www.cambridge-intl.com; www.cambridge-es.com

DURGESH INDUSTRIES

113, Jawahar Colony, N.I.T, Faridabad-121005, Haryana, India
www.wiremeshbelts.net

HEIGHTS WIRE BELT FACTORY

South of Development Zone, Dezhou, Shandong, China
www.wiremeshbelt.org

AB SANDVIK PROCESS SYSTEMS

2453, SE-81181 Sandviken, Sweden
www.processsytems.sandvik.com

STEINHAUS GMBH

Platanenallee 46, 45478 Mülheim an der Ruhr, Germany
www.bakingovenbelts.com

YANGZHOU JIANGDU HUADA METAL MESH BELT FACTORY

Sanfeng Industrial zone, Jiangdu City, Jiangsu Province, China
www.meshbelt.cn

YANGZHOU JINRUN MESH BELT MANUFACTURING CO.

No.2, Dachang Road, Daqiao Town, Jiangdu City, Yangzhou, Jiangsu, China, PC 22521
wmbbelt.com

INDEX

Note: Page numbers followed by "f" refer to figures.

Edwards Brothers Malloy
Ann Arbor MI. USA
August 23, 2016